Biochemistry Research Trends

Biochemistry Research Trends

The Biochemical Guide to Enzymes
David Aebisher, PhD
and Dorota Bartusik-Aebisher, PhD (Editors)
2022 ISBN: 979-8-88697-410-2 (Softcover)
2022 ISBN: 979-8-88697-518-5 (eBook)

Mineral Water: From Basic Research to Clinical Applications
Maria João Martins, PhD (Editor)
2022 ISBN: 978-1-68507-458-6 (Hardcover)
2022 ISBN: 978-1-68507-541-5 (eBook)

Terpenes and Terpenoids: Sources, Applications and Biological Significance
Charles A. Davies (Editor)
2022 ISBN: 978-1-68507-559-0 (Hardcover)
2022 ISBN: 978-1-68507-595-8 (eBook)

Circadian Rhythms and Their Importance
Rajeshwar P. Sinha, PhD (Editor)
2022 ISBN: 978-1-68507-547-7 (Hardcover)
2022 ISBN: 978-1-68507-585-9 (eBook)

A Biochemical View of Antioxidants
David Aebisher PhD, DSc,
and Dorota Bartusik-Aebisher, PhD (Editors)
2021 ISBN: 978-1-68507-151-6 (Hardcover)
2021 ISBN: 978-1-68507-295-7 (eBook)

More information about this series can be found at
https://novapublishers.com/product-category/series/biochemistry-research-trends/.

Santosh Kumar Karn and Anne Bhambri

Biomolecules and Corrosion

Copyright © 2023 by Nova Science Publishers, Inc.
DOI: 10.52305/VKVZ2636.

All rights reserved. No part of this book may be reproduced, stored in a retrieval system or transmitted in any form or by any means: electronic, electrostatic, magnetic, tape, mechanical photocopying, recording or otherwise without the written permission of the Publisher.

We have partnered with Copyright Clearance Center to make it easy for you to obtain permissions to reuse content from this publication. Simply navigate to this publication's page on Nova's website and locate the "Get Permission" button below the title description. This button is linked directly to the title's permission page on copyright.com. Alternatively, you can visit copyright.com and search by title, ISBN, or ISSN.

For further questions about using the service on copyright.com, please contact:
Copyright Clearance Center
Phone: +1-(978) 750-8400　　　　Fax: +1-(978) 750-4470　　　　E-mail: info@copyright.com.

NOTICE TO THE READER

The Publisher has taken reasonable care in the preparation of this book, but makes no expressed or implied warranty of any kind and assumes no responsibility for any errors or omissions. No liability is assumed for incidental or consequential damages in connection with or arising out of information contained in this book. The Publisher shall not be liable for any special, consequential, or exemplary damages resulting, in whole or in part, from the readers' use of, or reliance upon, this material. Any parts of this book based on government reports are so indicated and copyright is claimed for those parts to the extent applicable to compilations of such works.

Independent verification should be sought for any data, advice or recommendations contained in this book. In addition, no responsibility is assumed by the Publisher for any injury and/or damage to persons or property arising from any methods, products, instructions, ideas or otherwise contained in this publication.

This publication is designed to provide accurate and authoritative information with regard to the subject matter covered herein. It is sold with the clear understanding that the Publisher is not engaged in rendering legal or any other professional services. If legal or any other expert assistance is required, the services of a competent person should be sought. FROM A DECLARATION OF PARTICIPANTS JOINTLY ADOPTED BY A COMMITTEE OF THE AMERICAN BAR ASSOCIATION AND A COMMITTEE OF PUBLISHERS.

Additional color graphics may be available in the e-book version of this book.

Library of Congress Cataloging-in-Publication Data

ISBN: 979-8-88697-458-4

Published by Nova Science Publishers, Inc. † New York

Contents

Preface ix

Acknowledgments xi

Chapter 1 **Corrosion: An Overview** 1
  Abstract 1
  Introduction 1
  Microbiologically Influenced Corrosion (MIC) 3
  Mechanisms 5
  Economical Loss from the Corrosion 11

Chapter 2 **The Role of Biofilm in Corrosion** 13
  Abstract 13
  Introduction 13
  Biofilm in Industries and Their Significance 17
  Places Where Biofilm Form 18
  Organisms Involved in Biofilm Formation 19
  Mechanisms of Biofilm Formation 19
  Economical Loss Due to Biofilm Formation 21

Chapter 3 **Microbial Diversity and Their Importance** 23
  Abstract 23
  Introduction 23
  Microbial Diversity 24
  Microorganisms Involved in Corrosion Process 25
  Iron-Oxidizing Bacteria 26
  Iron-Reducing Bacteria 27
  Manganese-Oxidizing Bacteria 28
  Manganese-Reducing Bacteria 28
  Sulphur-Reducing or Oxidizing Bacteria 29

Chapter 4	**Techniques to Determine the Microbial Diversity**	**31**
	Abstract	31
	Introduction	31
	Morphological Characterization of Bacteria	32
	Biochemical Characterization	32
	Molecular Methods to Determine Microbial Diversity	33
Chapter 5	**Extracellular Polymeric Substances and Corrosion**	**49**
	Abstract	49
	Introduction	49
	Source of EPS	51
	EPS in Corrosion Process	54
	Techniques to Determine EPS	56
Chapter 6	**Proteins in Corrosion**	**57**
	Abstract	57
	Introduction	57
	Role of Proteins and Enzymes in the Corrosion Process	58
	Technique to Determine Enzymes in Corrosion	66
Chapter 7	**Role of Lipids in Corrosion**	**71**
	Abstract	71
	Introduction	71
	Role of Lipids in Corrosion Process	72
	Sample Preparation	74
	Separation and Analysis by Chromatography	75
	Lipid Fractions by TLC	75
	Fatty Acid Methyl Esters by GC	76
	Chemical Techniques	77
Chapter 8	**DNA or eDNA (Environmental DNA) in Corrosion Process**	**79**
	Abstract	79
	Introduction	79
	eDNA - A Source For Detection of New Species	80
	eDNA in the Corrosion Process	80

Chapter 9	**Technique to Identify Biomolecules**	
	or Biofilm in Corrosion	85
	Abstract	85
	Introduction	85
	Confocal Laser Scanning Microscopy (CLSM)	87
	Determination of Dry Weight and Total Carbohydrate Levels	88
	Analysis of Metal Ion Leaching and Accumulation by Biofilm	88
	Microscopic Analysis of the Biofilm	88
	Identification of Biofilm Bacteria	89
Chapter 10	**Omics Approach in Biocorrosion**	91
	Abstract	91
	Omics Approach	91
	Metabolomics: An Approach	93
	Metabolic Fingerprinting	94
	Metabolomics in Biocorrosion	94
	Analysis of Metabolic Foot Printing	95
References		97
Index		129
About the Authors		137

Preface

Corrosion causes great losses to the economy every year. Microbial-influenced corrosion (MIC) is important for maritime, chemical engineering, and bioprocess engineering industries. Presently no environmentally friendly technology has been available to minimize the economic loss of biocorrosion. Currently available anti-biocorrosion technology depends heavily on chemical methods to regulate biofilm formation which has a negative impact on the environment. Therefore, it is essential to know the fundamentals of the roles of biomolecules within the whole process to develop detailed research capabilities and potential control and management strategies. This book targets the roles of EPS, proteins, lipids, DNA, and different metabolites currently known to be involved in the corrosion processes. The potential roles of EPS, proteins, lipids and enzymes are still poorly understood. There are still collective issues that need to be addressed, including the importance of the microbial role in MIC. More specifically, there exists a need to understand the impact of enzyme activities inside the biofilm matrix on the dynamics of corrosion reactions. Also, there is the involvement of metals or organometallic complexes in electron transfer and from chemically and morphologically diverse metallic surface films to ultimate electron acceptors.

Chapter 1: The aim of this work is to summarize microbiologically influenced corrosion which is a challenging issue and is conducted on steels, biocide enhancers, antibacterial stainless steels, and antibacterial coatings for mitigation.

Chapter 2: In the present and next chapter, the authors report on the various organisms which are involved in the formation of biofilm and mechanism and the regulation of biofilm production.

Chapter 3: Here, the authors report on microbial diversity and the microorganisms that are involved in the corrosion process such as iron-oxidizing bacteria and iron-reducing bacteria.

Chapter 4: In this chapter, the authors report on the various techniques to decipher the diversity of organisms such as morphological characterization, biochemical characterization and molecular characterization.

Chapter 5: This chapter reports on the various techniques to characterize the exopolysaccharide (EPS), which consists of different macromolecules, which mediate initial cell attachment to the material surface and constitute a biofilm matrix.

Chapter 6: This chapter reports on certain types of interactions between the substrates and cells and the enzymes in the corrosion process.

Chapter 7: This chapter targets the roles of lipids and different substances currently known to be involved in corrosion processes.

Chapter 8: This chapter describes the role of eDNA in the corrosion process observed in specific metals due to surface charge density and passivation conditions.

Chapter 9: This chapter discusses various techniques that are used to identify the biofilm in the corrosion process.

Chapter 10: This chapter discusses genomics, proteomics, and transcriptomics to understand the biological mechanisms of biofilm formation and the corrosion process.

Acknowledgments

The authors are thankful to Gaurav Deep Singh, Chancellor, Sardar Bhagwan Singh University, Balawala, Dehradun, Uttarakhand, India, for providing the facility, space and resources to conduct this work successfully. Professors Karn and Bhambri would like to acknowledge the publisher, editors, and support staff at Nova Science Publishers as they were very helpful in the various stages of developing and producing the book.

Chapter 1

Corrosion: An Overview

Abstract

Corrosion is an electrochemical process which mainly occurs due to the transfer of electrons. Metal corrosion occurs because of the biochemical activity on the surface of the metal or other factors such as secretion of enzymes, biofilm development, exo-polymeric substance production and metal destruction due to bacteria. There are various microbes mentioned such as sulphur-oxidizing bacteria, iron reducers, manganese oxidizers and sulphate-reducing bacteria that are associated with the steel or iron corrosion. Microbiologically influenced corrosion is the most challenging issue and conducted on steels, biocide enhancers, biocides, antibacterial stainless steels and antibacterial coatings for the mitigation. They play an important role in the corrosion of various materials as well as in the degradation such as magnesium, carbon steel, aluminium alloy and dulex stainless steel. Several different mechanisms have been mentioned such as cathodic depolarization by hydrogenase, volatile compound of phosphorus, anodic depolarization, Fe-binding exopolymers, sulphide, biocatalytic cathodic sulphate reduction. Corrosion has been estimated to cause economic loss to the tune of about 4 trillion US dollar. This chapter discusses in detail corrosion, biocorrosion and their mechanisms.

Keywords: corrosion, biocorrosion, metal, carbon steel, stainless steel

Introduction

Corrosion is a worldwide issue which affects several municipal services as well as industries like oil refinery, drinking water systems, shipping, up keep of statues and historical buildings and constructions (Warscheid and Braams, 2000; Videla and Herrera, 2005). Corrosion is the amalgam of physical, chemical and microbiological processes leads to the degradation of materials like steel, iron, stone, concrete and metal as well as alloys. This happens due to chemical, electrochemical and biochemical interactions in between metals and alloys. Moreover, it also releases the metal ions into the environment

(Kip and Veen, 2015; Warscheid and Braams, 2000; Videla and Herrera, 2005). Corrosion is basically an electrochemical phenomenon which usually happens because of the mechanism of electron transfer in the presence of an electrolyte in which oxygen acts as an ultimate electron acceptor, while protons also may play a vital role as an electron acceptor in some cases (Beech and Sunner, 2004; Coetser and Cloete, 2005).

Oxygen, temperature, moisture, chlorides, organic as well as inorganic acids and high pressure forms an integral part of corrosive environments. At the time of corrosion, metals get converted to more thermodynamically stable compounds like carbonates, oxides, salts and hydroxides. By spontaneous corrosion, the recovering of compounds such as minerals and ores from the metal, results in reduction of in free energy. Consequently, during the corrosion reactions, the energy which is used for obtaining metal from alloying or ore gets released (Veronika, 2008; Fontana, 1986). Metal corrosion occurs due to bio-chemical erosion from the metallic surface. While anodic reactions are stimulated by corrosion chemicals such as acids, cathodic reactions are induced by the microorganisms which consume hydrogen and secrete acidic metabolites as well as enzymes (Videla and Herrera, 2005).

Biocorrosion occurs particularly due to electrochemical reactions driven or influenced by biofilm forming microbes. Biocorrosion or microbiologically influenced corrosion was recognized as an essential category of the corrosion almost 50 years ago. Microbes affect destruction of metals by several means like enzymatic digestion which promotes the erosion processes at cathodic sites, development of biofilm or the production of exopolymeric substances which changes the conditions at metal surfaces, produces the metabolic products of corrosion towards the metal or the protective layer and deterioration of chemical compounds which increase or inhibits the corrosion (Coester and Cloete, 2005; Hiibel et al., 2008).

Lately, it became apparent that microorganisms not only cause corrosion but also protects against corrosion (Zuo, 2007). Since early 1990s, the role of microbes on metal corrosion has been known (Kuehr et al., 1934; Videla and Herrera, 2005). In the gas and oil industry, the microbiologically influenced corrosion is one of the biggest challenges. In oil field, the pitting attacks of microbiologically influenced corrosion results in equipment, pipeline failures and reservoir souring. For such cases of corrosion, sulphate-reducing bacteria are always responsible (Lee et al., 1995; Hamilton, 1985; Cord-Ruwisch et al., 1987).

These sulphate-reducing bacteria are anaerobic as well as non-pathogenic but in the reduction reaction of sulphate to sulphide it also acts as a catalyst (Kakooei et al., 2012) which means they are able to form severe. There are several bacteria which are associated with corrosion of steel as well as iron, manganese oxidizers, iron reducers and sulphate-reducing bacteria are such microorganisms which secretes organic acids as well as the extracellular polymeric substances (Hamilton, 1985; Zuo, 2007).

Microbiologically Influenced Corrosion (MIC)

In industrial systems, microorganisms are present everywhere. In gas as well as in oil industry, the microbiologically influenced corrosion is one of the most challenging issues (Brooks, 2013; Javaherdashti, 2016; Koch et al., 2018). Over the past two decades, these pipelines facility failures and leakages because of microbiologically influenced corrosion occurs commonly leading to damages in environment (Bhat et al., 2011; Hinkson et al., 2013; Jacobson, 2007).

To mitigate the microbiologically influenced corrosion, research on antibacterial stainless steels, biocides, antibacterial coatings and biocide enhancers have been conducted (Li et al., 2016; Qian et al., 2017; Li et al., 2017; Xu et al., 2017). Because of the impending need to recover more and more oil, microbiologically influenced corrosion becomes more prevalent as it leads to reservoir souring (Xu et al., 2013). Moreover, as the awareness towards microbiologically influenced corrosion increases, it leads to further identification of more such issues.

Microbial corrosion is associated with the enhanced maintenance costs in power plants, production loss and also expensive equipment damages (Maria et al., 2014). The infestation by microorganisms causes corrosion and biofouling to infrastructures and equipment in various industrial settings such as cooling water systems (Berndt, 2011; Jia et al., 2017), water storage facilities of nuclear (Dai et al., 2016); water distribution systems (Teng et al., 2008), medical devices (Jia et al., 2017), underground storage tanks (Sowards and Mansfield, 2014), rail systems (Maruthamuthu et al., 2013), ships (Heyer et al., 2013). Microorganisms play a vital role in the degradation as well as in corrosion of various materials such as stainless steel (SS) (Zhang et al., 2015); high nitrogen (HNS) (Li et al., 2016); carbon steel (Xu et al., 2017); super austenitic SS (SASS) (Li et al., 2017); aluminium alloy (Dai et al., 2016); magnesium (Ahmadkhaniha et al., 2016);

duplex SS (DSS) (Li et al., 2017; Xia et al., 2015; Xu et al., 2017; Xu et al., 2017; Zhou et al., 2018); hyper dulex SS (HDSS) (Li et al., 2016) and even concrete (Harbulakova et al., 2013). It is notable that microbiologically influenced corrosion itself may not cause complete manifestations of corrosion, rather, it interacts with other processes of corrosion such as under-deposit corrosion (Wang et al., 2017), fatigue crack tip embrittlement (Sowards et al., 2014) and stress corrosion cracking (SCC) (Wu et al., 2015). Figure 1 represents various types of corrosion.

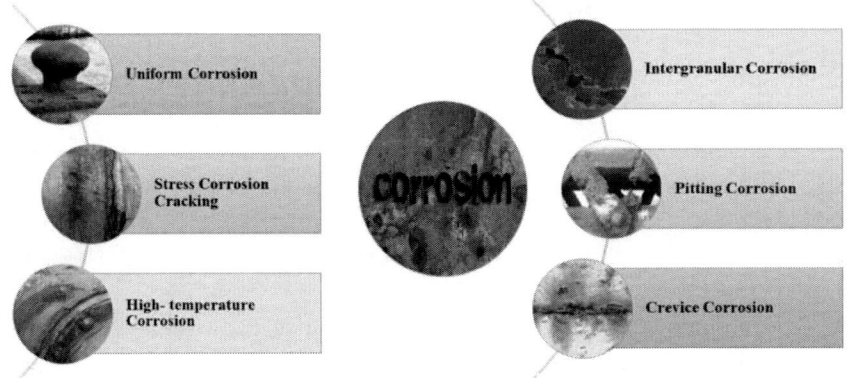

Figure 1. Various types of corrosion.

Biocorrosion has become major concern for corrosion scientists and engineers from different fields across the globe. There are various microorganisms that are capable of inducing corrosion belonging to diverse kingdoms like fungi, bacteria and archaea. It has been found that the sulphate-reducing archaea and bacteria are the major causative organisms for decades as sulphate is well distributed in such ecological systems as agricultural runoff, sea water and brackish water (Li et al., 2016; Jia et al., 2017).

Frequently, these sulphate-reducing archaea and sulphate-reducing bacteria were isolated from those gas and oil industries complaining of microbiologically influenced corrosion (Javaherdashti, 2016; San et al., 2012). According to Aktas et al., 2017, it has been observed that in the presence of sulphate-reducing bacteria, there is a positive correlation between the sulphide generation, sulphate consumption and pitting corrosion. Additionally, there are other microorganisms that are linked to corrosion; these include manganese-oxidizing bacteria (Ashassi-Sorkhabi et al., 2012), nitrate-reducing bacteria (Xu et al., 2013), diverse fungal species (Ching et al., 2016), iron-oxidizing bacteria (Chandrasatheesh and Jyapriya, 2014) and methanogens (Aktas et al., 2017; Tan et al., 2017). There are some species of microorganisms that forms biofilms causing the microbiologically influenced corrosion such as *Cytophagajohnsonae, Pseudomonas, Sphingomonas, Rhodotorula, Acidovoraxdelafieldii, Pseudomonas paucimobilis, Flavobacterium, Acidovorax* and *Micrococcus kristinae* (Angell and Chamberlain, 1991; Arens et al., 1995; Critchley and Fallowfield, 2001; Critchley et al., 2004).

In the field of metal corrosion, this microbiologically influenced corrosion is an essential branch with much scope in research. Jacobson, (2007) pointed out that this microbiologically influenced corrosion was the main reason behind the leakage case of Alaska oil pipeline at Prudhoe Bay in 2006. Globally, the prices of crude oil cause a major spite and drawn considerable attention of public to potential damages in the environment such as the leakage of pipeline (Walsh et al., 1993).

Mechanisms

There are several different mechanisms that has been proposed to explain the phenomenon of biocorrosion. According to von Wolzogen Kuehret al., (1934) the cells of sulphate-reducing bacteria use hydrogenase enzymes for lowering the activation energy of hydrogen atom and also the desorption process formulated in the classical cathodic depolarization theory. This is one of the rate-limiting steps for microbiologically influenced corrosion by sulphate-reducing bacteria. Microbial corrosion is thus, explained by the cathodic depolarization theory that is caused by hydrogenase-positive sulphate-reducing bacteria. Nevertheless, microbially influenced corrosion is not well understood due to the absence of clear explanation of bio-electrochemical process that occurs at the interface between metal matrix and biofilm.

Cathodic Depolarization by Hydrogenase

Cathodic depolarization mechanism of corrosion was proposed by Kuehr and Vlugt, (1934) induced by sulphate-reducing bacteria that get depolarized via the oxidation of cathodic hydrogen as articulated in the cathodic depolarization theory. When metal surface is exposed to water, it becomes polarized on losing positive metal ions (anodic reaction). The free electrons reduce water-derived protons (cathodic reaction) in the absence of oxygen, to produce hydrogen that inhabits on the metal surface and establish a dynamic equilibrium. It is expected that the hydrogen thus produced is consumed by the sulphate-reducing bacteria (Marcus, 2002; Castaneda and Benetton, 2008). The anodic dissolution of metal gets enhanced in this mechanism and therefore corrosion products such as Fe $(OH)_2$ as well as FeS are formed (Costello, 1974).

Iron Sulphides (King's Mechanism)

King's mechanism explained the formation of solid FeS on the metal surface that plays an important role of absorber of molecular hydrogen (king and JDA, 1971). The area that is covered by iron sulphide becomes cathodic at this state where as the area of biofilm behaves as anodic (King and JDA, 1971; Videla, 2000; Hilbert et al., 2005). The crack-up of the protective Mackinawite film (Mackinawite is a layer that is formed of iron sulphide and can form instantly at temperature under 100°C in relatively a short period within the sour corrosion systems leads to enhancement of Fe concentration in the solution and when this film gets broken, then the rate of corrosion accelerates in terms of Fe concentration (King and J. D. A, 1973). It has been found that there is no sulphide film formed after enough FeS formation for initiating a galvanic cell among Fe as well as FeS whereas due to galvanic corrosion a high corrosion rate was recorded (King and JDA, 1971). In an anaerobic reactor of biofilm, the impact of suspended FeS corrosion on the mild steel corrosion has been studied where the concentration of Fe^{2+} rises from 0 to 60 mg/l (Lee and Characklis, 1993). When the concentration of Fe^{2+} reaches up to 60 mg/l then the particles of FeS penetrate via the protective film of iron sulphide, and the protective film gets broken. Lee and Characklis (1993) demonstrated in the SEM imaging the intergranular corrosion detected on the surface of metal.

Volatile Compound of Phosphorous

According to Iverson and Olson (1983), phosphorous compound which is a volatile material is also responsible for the event of corrosion. They also asserted that the reducers of sulphate accelerate corrosion via the production of tremendously corrosive phosphorous compounds like phosphine that leads to the formation of iron phosphide (Iverson, 1968; Iverson and Olson, 1983). It has been shown that the compounds of phosphorus found in the extracts of yeast seem to be a predecessor to the corrosive phosphorous compound (Iverson, 2001).

Anodic Depolarization

Iron-reducing bacteria that cause anodic depolarization have become the subject of extensive investigation. In an aqueous anaerobic environment, the corrosion of iron is an electrochemical event in which electrons get generated via anodic reaction of metal while hydrogen is generated during the dissociation of water. With the combination of these two reactions molecular H_2 is generated and this process is known as cathodic polarization in which a layer of hydrogen may protects both metal surfaces. The main product of corrosion is $Fe(OH)_2$. When sulphate-reducing bacteria reduces sulphate to sulphide consumption of hydrogen take place. And then the dissociation of H_2S enhances the concentration of hydrogen in cathodic area and changes the kinetics. Further, a new corrosion product of FeS gets formed with the anodic depolarization (Obuekwe et al., 1981; Crolet, 1992; Araujo-Jorge et al., 1992; Ford and Mitchell, 1990; Coetser and Cloete, 2005; Wang and Liang, 2007). Next, via the anodic depolarization process of sulphide, sulphate-reducing bacteria accelerates the anodic active dissolution of 1OCrMoAI steel in seawater. The sulphate ion that forms from the sulphate-reducing bacteria activity reacts with Fe_2^+ ions and forms FeS which accelerates the anodic active dissolution (Wang and Liang, 2007).

Fe-Binding Exopolymers

Normally the biological material that deposits on any surface is known as biofouling or biofilm. Biofilms also known as a community of bacteria as well as extracellular polymer substance such as polysaccharides. The main purpose of biofilm formation is to protect the bacteria and to trap the nutrients for growing bacteria. The role of biofilm also influences in various industries like oil field industries (Cord-Ruwisch et al., 1987), drinking water system industry (Momba, et al., 2000) as well as food and dairy industry

(Kumar and Anand, 1998). Extracellular polymeric substances (EPS) that are produced by sulphate-reducing bacteria have the ability to accelerate the corrosion by binding with the metal ions (Beech and Cheung, 1995; Beech et al., 1996; Beech, 2004; Fang, 2002). Different rates of corrosion have been shown by sulphate-reducing bacteria with extracellular polymeric substances of various composition. According to Beech et al., (1998) and Beech et al., (1994) Uronic acid has been detected when extracellular polymeric substances get released by a relatively aggressive *Desulfovibrio* strain whereas according to Fang et al., (2000) the production of extracellular polymeric substances increases in the presence of Cr and accelerate the corrosion of mild steel in seawater (Fang et al., 2000). Chan et al., (2002) confirmed that extracellular polymeric substances is the lone agent of metal corrosion. On the surface, biofilm forms in four stages such as growth within the biofilm, transportation of bacterial cells from bulk to surface, attachment of bacterial cells and the transportation of organic material to metal surface (Videla, 2001). Due to the accumulation of biofilm, these mechanisms modify the interface structure which is considered as the main reason of microbially influenced corrosion.

Sulphide and Hydrogen-Induced Stress Corrosion Cracking (SCC)

Hydrogen embrittlement as well as corrosion-fatigue crack growth enhances the activity of sulphate reducing bacteria (Videla et al., 2005; Edyvean, 1998). In natural water of sea, the biologically produced hydrogen sulphide accelerates the corrosion fatigue crack growth of the high strength steel (Dom alicki et al., 2007). Due to the bacterial activities and their interactions with other components of the environment like the degradation by large fouling organisms, production of extracellular polymeric substances as well as the interactions of metal and microbes. The surface of metal is surrounded by two different local environments with or without the microbes even with the same levels of sulphide. Moreover, due to the high levels of hydrogen, the sour environments are precisely corrosive at the metal surface or in a crack due to the activation of sulphide at the cathode (Videla, 2000).

Due to the presence of organic molecules on the extracellular polymeric substance's matrix as well as on the metal surface, the influence of hydrogen can be modified that describes the differences between general embrittlement effects (as measured by hydrogen flux) and crack tip effects (as measured by crack growth) (Videla et al., 2005). The differences between biotic and abiotic solution having similar levels of corrosive compounds like iron sulphides that can be related with the heterogeneities that produce at the surface of metal by the formation of biofilm in the presence of extracellular polymeric substances (Videla, 2000). According to Domalicki et al., (2007) at the low as well as medium cathodic polarization, sulphate reducing bacteria may produce hydrogen sulphide. Hydrogen sulphides inhibit the formation of deposits and reduce the pH near electrode electrolyte, therefore, encourages the loss of plasticity as well as hydrogen charging and these effects are used in the study of steels function. According to (Videla, 2000), the same amount of hydrogen makes a less detrimental effect on the sorbite steel of increased strength, but at similar cathodic polarization, this steel absorbs the highest amount of hydrogen and reveals the most pronounced degradation.

Role of Sulphides

In microbiologically induced corrosion, the role of biomineralization has been investigated by Little et al., (1998). Due to the metal-reducing bacteria, the biomineral dissolution reaction either removes the oxide layers or forces the mineral replacement reaction which helps in metal decomposition. The deposition of mineral on a metal surface shifts the potential of corrosion either in a positive or negative direction depending up on the nature of mineral and is known as biomineralization. Bio-precipitated sulphides move the corrosion potential in a negative, more active direction, resulting in accelerated corrosion of some metals and alloys. Iron oxide formation can begin a sequence of events that results in under deposit corrosion of susceptible metals (Little et al., 1998; Little et al., 2000).

Three Stages Mechanism (Romero Mechanism)

Lately, a three-stage mechanism has been proposed by Romero for the sulphate-reducing bacterium which induces corrosion of iron (Romero, 2005). In the first stage, adsorptions of iron sulphide product as well as the microbial cells occur. Further, via the iron sulphide products, a micro galvanic corrosion cells are formed (Mackinawite and Pyrite) as well as the metallic surface which generates hydrogen permeation peak (Romero, 2005; Ocando et al., 2007).

In the second stage of equilibrium between inorganics and microbial cells, the metal gets somewhat ennobled due to the development of combination film of a denser extracellular polymeric substances as well as iron sulphide film (Romero, 2005; Ocando et al., 2007). The two major occurrences that take place are the stabilization of film and microbial corrosion. In the third stage, the local pH gets reduced which causes the activity of sulphate-reducing bacteria on the steel in the presence of hydrogen sulphide. Locally, these sulphate-reducing bacteria decreases the Pyrite to Mackinawite and produces a severe localized corrosive process which configured into the groups of rounded and deep holes and then detachment happens. A galvanic corrosion is supposed to be generated between the cathodic iron sulphide products as well as the anodic metal composed of Greigite, Troilite, Esmitite, Mackinawite, Pyrrhotite, Marcasite, and Pyrite. Due to the anti-diffusive effect of the extracellular polymeric substances or the barrier, no hydrogen permeation happens in this more aggressive phase. In this stage, the microbes grow exponentially to about 10CFU/cm^2 that generates sufficient hydrogen sulphide. Thus, without the atomic hydrogen absorption, corrosion gets accelerated.

Biocatalytic Cathodic Sulphate Reduction (BCSR)

In the biocatalytic cathodic sulphate reduction theory, MIC takes place since the sulphate reduction at the cathode will consume the electrons released by iron dissolution at the anode with the help of biocatalyst and the interface of biofilm and the metal are place for both anodic and cathodic sites (Gu et al., 2009; Zhao, 2009). It assumes that a corrosive sulphate-reducing bacteria (SRB) biofilm is formed on an iron surface causing the following reactions to go forward due to biocatalysts.

Cathodic reaction shows the half reaction of reduction from sulphate to sulphide due to biofilm catalysis. Some species were added solely to balance the charges and elements in order to be consistent with other reactions. One should not interpret cathodic reaction strictly in terms of converting proton to hydroxide because the actual sulphate reduction in SRB is coupled with other biochemical reactions. An increase in sessile SRB population may be observed due to externally supplied electrons in impressed current cathodic protection situation. Another factor may be that SRB cell wall carry charges that are attracted to the surface. If there is a stoppage of electron supply, the sessile SRB cells may turn to attacking iron to get electrons for BCSR. To assure a steady supply of externally supplied electrons, a continuous impressed current is desired. A more negative voltage (i.e., a larger driving force) is needed to deliver the current due to the increased ohmic resistance exerted by the biofilm (Gu and Xu, 2010).

Economical Loss from Corrosion

The gross national loss for the developed countries by corrosion is projected between 1 and 4% (Mehana, 2009). According to Fleming et al., (1996), it has been estimated that the problem of corrosion is due to the presence of microorganisms. Additionally, Jack et al., (1992) estimated the oil companies experiences 34% of corrosion related damages.

The microbiologically influenced corrosion causes a major spike in the global crude oil prices, and due to the leakage of pipelines inflicted by biocorrosion, the considerable attention has been drawn towards a potential damage to the environment. According to Walsh et al., (1993), it has been found that around 20% or more of all corrosion can be attributed to microbiologically influenced corrosion amounting to billions of dollars getting drained each year in the United States alone.

Due to the leakage of these pipelines, it not only causes financial loss but also causes harm to environment. It has been described that the release of highly toxic gases like H_2S or flammable liquids as well as gases are the major safety concerns. It has also been found that oxygen being highly corrosive gets removed from the pipelines by using oxygen scavenger. Nevertheless, this anaerobic corrosion leftovers are a severe threat. There are some examples of conventional chemical corrosion which threaten the industry of pipeline such as acetic acid corrosion, H_2S corrosion and CO_2 corrosion (Gu and Xu, 2013).

There are some oxidants which do not cause corrosion abiotically like sulphates as their reduction involves bio-catalysis. When corrosive biofilms are present, they turn out to be hazardous. Sulphate-reducing bacteria are the most common bacteria associated with anaerobic microbiologically influenced corrosion because of the wide availability of sulphate in various aqueous environments such as seawater that is typically used in water injection to increase oil reservoir pressures (Boopathy et al., 2002; Javaherdashti, 1999). These sulphate-reducing bacteria are an anaerobic bacterium which uses sulphate as electron acceptor for the production of hydrogen sulphide (Liamleam and Annachhatre, 2007) and without any growth these bacteria can tolerate exposure to oxygen for a period of time (Thauer et al., 2007).

In different industrial sectors, the economic costs of corrosion of iron and their alloys like gas and oil operations are well recognized (Borenshtein, 1994; Videla, 1996). Collectively, corrosion also causes damage in marine steel infrastructure like pipeline systems as well as offshore oil installations which also lead to revenue losses.

Koch et al., (2002) and Flemming, (1996) demonstrated that the economic cost of corrosion is almost 4% of the GDP of industrialized countries. Moreover, it has also been estimated that approximately 20% of its cost is on account of microbial activity (Heitz et al., 1996; Beech and Sunner, 2004). In anaerobic corrosion, the interference of sulphate reducing bacteria is now well accepted (Beech, 2004; King and Miller, 1971; Castaneda and Benetton, 2008; Sheng et al., 2007; Miranda et al., 2006; Santana Rodriguez et al., 2006; Ilhan-Sungur et al., 2007). It has also been found that not only the sulphate reducing bacteria causes anaerobic corrosion, instead, biocorrosion can also occur in the absence of sulphate reducing bacteria (Lopez et al., 2006). Yearly, corrosion causes 4 trillion USD financial losses worldwide, half of which is because of damages by corrosion whereas other half is due to the cost of corrosion protection measures (Hou et al., 2017; Li et al., 2015).

Chapter 2

The Role of Biofilm in Corrosion

Abstract

A complex structure of microbiome that have various microbial colonies or single type of cells in a group are the biofilms that attach to the surface and these cells get embedded in extracellular polymeric coat i.e., a matrix that is mainly composed of polysaccharides, eDNA and proteins. There are various artificial substances that are developed by the biofilm common in wood, polyethylene, food, stainless steel, glass and polypropylene industry etc. These biofilms releases toxins which contaminate and causes multiple intoxications as well as material deterioration.

There are various organisms which are involved in the formation biofilm such as *Listeria monocytogenes, Salmonella enterica, Escherichia coli, Bacillus cereus, Geobacillus Stearothermophilus* and *Pseudomonas* sp. Biofilms form via attachment, dispersion and maturation the three stages in the formation of biofilm. Biofilm-embedded sessile community having heterogeneous cells have a wide range of different tools to withstand the antimicrobials. The mechanism and the regulation of biofilm production are by quorum-sensing system, inactivation of biofilm formation and the development of resistance patterns of biofilm-embedded microorganism against antimicrobials. These bacteria pose serious health problem for various industries like medical, water, power plant, marine, dairy, water logging hampers, food and pipe blockages that decreases the economic loss.

Keywords: biofilm, surface, environment, aqueous layer, colonized organism

Introduction

The existing aquatic environment was immensely oligotrophic millions of years ago when microbes were the only life forms on earth. Due to the hostile environmental factors such as acidity, UV radiation and heat, the niches permissive were limited for life (Costerton and Lappin-Scott, 1995). Until the niche permissive was found for the growth, the purpose of free-living mode of microbial growth and planktonic enabled them to move from one habitat to another. The formation of biofilm allowed these sessile organisms to continue inhabiting in the places as well as utilize and trap the scarce organic compounds. Among the microbial groups, the development of co-operation permits the use of more refractory as well as more complex nutrients. The environment of colonized surface changed due to formation of biofilm and subsequently became more suitable for the growth of bacteria. For bacteria, the biofilm formation became the means of survival and also became the reason why microbial biofilms are the characteristics of life forms that are found in extreme environments such as rock surface in oligotrophic mountain streams, hot springs as well as in cold. Currently, in industrial systems like breweries (Storgårds, 2000), drinking water distribution networks, (Szewzyk et al., 2000; LeChevallier, 1999; Koskinen et al., 2000), power plants (Linhardt, 1996), dairies (Pirttijärvi et al., 1998) and paper machines (Väisänen et al., 1994, 1998; Claus and Müller, 1996) the same survival strategy has been used successfully by the microbes. The formation of biofilms causes several technical problems such as it enhances the flow resistance of pipe lines, reduction in the thermal transfer capacity in heat exchangers as well as contamination of process of drinking water by microorganisms (Flemming, 1996). The presence of pathogenic bacteria in the form of biofilms in the drinking water distribution system is an important risk factor for public health (*Legionella pneumoniae*, *Klebsiella pneumoniae*, *Mycobacterium avium*) (LeChevallier, 1991 and 1999; Szewzyk et al., 2000).

In an aqueous environment, the non-living surfaces promptly gather inorganic ions as well as organic molecules to form a layer known as conditioning film. Consequently, the planktonic microbes attach on the surface of a film rather than the non-living surface which may have various chemical properties in all but mainly in the most oligotrophic ecosystems (Costerton and Lappin-Scott, 1995; Marshall, 1997). Initially the attachment of microbes on the surface often involves a flagellum or exopolysaccharide (EPS) as well as a portion of the cell whereas the microbial cell continues to revolve (Marshall, 1997). While searching a suitable location, microbes use its motility to sustain contact with the surface during the reversible attachment (Korber et al., 1995). This search known as chemo-sensoring when microbes prefer the definite substrates present on the surface or produced by another microbe (Nielsen et al., 2000). The method of surface translocation such as flagellar motility of microbes is known as positioning mechanisms (Davey and O'Toole, 2000; Martínez et al., 1999; McBride, 2001).

In *Escherichia coli*, the flagellum-mediated motility is essential for both moving and approaching across the surface. It has also been found that in *E. coli*, the outer membrane proteins and type I pili are essential for a stable organism surface interaction. According to Davey and O'Toole, (2000) *Pseudomonas aeruginosa* uses their flagellum only for taking the cell into the proximity of a surface. Initially reversible attachment can be transformed into an irreversible one or to detachment of the microbial cell. Naturally, the reversible attachment is considered to be predominant (Korber et al., 1995). On the basis of interaction between the substratum and the cell, the initial attachment of a single bacterium has been done. Coaggregation is the process in which the maturation of the biofilm and subsequent growth depends upon the cell-to-cell interactions as well as also known for the adhesion and recognition between genetically distinct bacteria (Whittaker et al., 1996). Coaggregation and the cluster of biofilms has been formed and regulated by means of signaling systems between the cells that shares the same location irrespective of whether they are related or not (Xie et al., 2000).

Rapidly biofilm get colonized and adhere strongly on metal surfaces which are in contact with seawater together with other organic matter that adsorbed and dispersed, resulting in a big complex microfouling, corrosion products and algae and other microorganisms (Fox et al., 2015). The concentration of chloride or other aggressive anions under the biofilms as well as corrosive active compounds or that produced in the environment by microorganisms like organic acids, acidic polysaccharides, dissolved CO_2, reduced pH, anaerobic sulphide and leads to the disruption of passive layer that protects the surface of metal (Wagner et al., 2004). Extracellular polymeric substances such as copper binding proteins can accelerate copper alloys as well as corrosion of copper that secreted by biofilm (Usher et al., 2014). Extracellular polymeric substances act as binding agents for copper ions which inhibits the growth of microbes. Consequently, electrochemical characteristics of metals surface gets modifies by biofilm (Mollica, 1992; Huttunen-Saarivirta et al., 2018) and causes physical deterioration or degradation which known as bio-corrosion or microbiologically influenced corrosion (MIC) (Beech, 2004).

According to Dunne, (2002) biofilms are the interface-associated colonies of the microorganisms which are embedded into a self-produced organic polymeric matrix and contain molecules that derived from the corrosion products or bulk aqueous phase of the metal substratum. It has also been found that in the nature, vast majority of bacterial life exists in the form of biofilms. Consequently, the formation of biofilm viewed as a universal bacterial survival strategy (Beech, 1995; Costerton et al., 1987; Costerton et al., 1995). The different types of biofilm matrix molecules such as lipids, proteins, DNA, polysaccharides etc. known as extracellular polymeric substances which plays an important role in their nutrition and structural integrity. According to Karatan and Watnick, (2009) at different stages of biofilm maturation, each of the exo-polymeric components has a comparative importance. The eternal cell adhesion is mediated by the combination of extracellular polymeric substances as the electrostatic interactions contributes to primary and reversible bacterial attachment to the surface that promotes strong binding of the microbes to the surface as well as with each other. Extracellular polymeric substances provide the anchoring sites for the extracellular enzymes that helps to degrade the inhibitory compounds or to access the various different substances and also responsible for the retention of water (Flemming and Wingender, 2010). Around the biofilm organisms these extracellular polymeric substances maintain an extremely hydrated micro-environment and also in water-deficient environments it enhances their tolerance to desiccation. The compounds of extracellular polymeric substances are an essential factor of metal-bacterial interactions in the case of corrosion-associated biofilms that contributes to the early development of biofilm.

Microorganisms especially bacteria are responsible for the composition of microbial biofilm which gets embedded in a heterogenous matrix of extracellular polymeric substances which help in the attachment of bacterial community that grows in aggressive environment and exists at the metal-liquid interface (Figure 2). Extracellular polymeric substances are the complex mixture of several macromolecules such as amphiphilic polymers, proteins, lipids, nucleic acids and polysaccharides whose nature depends upon the specific bacterial species in the biofilm as well as also on the relevant environmental factors like environmental stress and nutrient availability (Flemming and Wingender, 2010). The capability of extracellular polymeric substances determines the reactivity of the biofilm to copper as well as the microbial cells itself to bind with copper ions (sorption) and retain the copper particles to produce as a consequence of the corrosion by-products precipitation (Beech and Sunner, 2004; Edwards and Sprague, 2001). This patchwork of solid corrosion by-products, extracellular polymeric substances as well as microbes forms a complex reactive surface which induces the metallic copper oxidation, resulting in the release of corrosion by-products as well as ions into the water throughout the operational flow-stagnation cycle.

The process of accumulation done due to the formation of biofilms on metal and not essentially in uniform space or time which starts instantly after the immersion of metal in the aqueous environment (Characklis and Marshall, 1990). In a first stage a thin film of approximately 20-80 nm thick gets formed because of the deposition of organic compounds of high relative molecular mass and inorganic ions. This initial film alters the wettability of the metal surface as well as electrostatic charges which further facilitates its colonization by microbes. In a short time (hours or minutes depends upon the aqueous environment in which the metal gets immersed), the production of EPS and the growth of bacteria results in the biofilm development. Biofilm is different transport processes and chemical reactions occur at the interface of bio fouled that takes place via the thickness of biofilm (Characklis, 1981; Karn et al., 2020).

On a material surface, the attachment of microbes as well as subsequent formation of biofilm occurs once it comes in contact with a non-sterile aqueous environment (Costerton et al., 1987). Physical and chemical properties of these materials are colonized by the activity of microbes on the metallic surface where biofilms get forms.

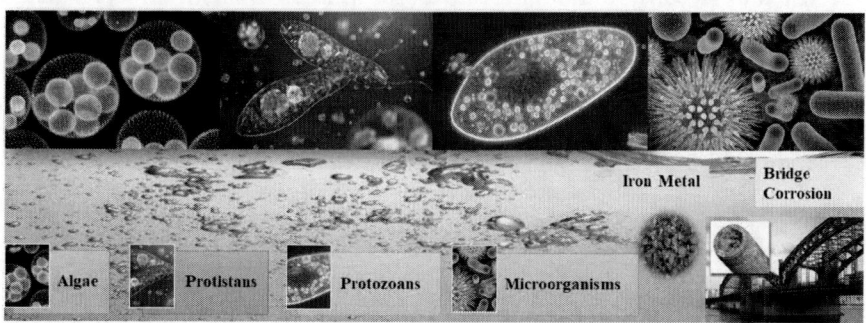

Figure 2. Interrelation among the organisms during biofilm formation on the surfaces.

Biofilm in Industries and Their Significance

The complex bacterial ecosystems of biofilms are formed by one or more species that get absorbed in an extracellular matrix having different compositions depends upon the type of colonizing species as well as on the industrial setup. Fungi and bacteria are the predominant microorganisms which may comprise these biofilms. As more than one bacterial species facilitate the attachment of biofilm to the surface, its presence has many essential ecological advantages. According to Meyer, (2015) mixed biofilms show higher resistance to disinfectants like quaternary ammonium compounds as well as other biocides.

Biofilms can form rapidly in the environment prevailing in food industry. Initially, the conditioning of the material surface as well as the reversible binding of the cells to that surface occurs. Further, binding becomes irreversible and then the microcolonies begin to develop. Finally, the three-dimensional structure of biofilm is produced giving rise to a complex ecosystem that is ready for dispersion (Nikolaev and Plakunov, 2007; Srey et al., 2013; Coughlan et al., 2016).

In food-based factories, biofilm-forming species are generally human pathogens. Due to these pathogens, the structures of biofilm get develop on different artificial substrates which are common in food industry such as wood, polypropylene, stainless steel, rubber, polyethylene, glass industry etc. (Abdallah et al., 2014; Colagiorgi et al., 2017).

In some industries like dairy, where several structures and processes act as surface substrates for the formation of biofilm with different colonizing species and at different temperatures, biofilm-associated effects such as alteration of organoleptic properties due to secretion of proteases or lipases, pathogenicity and corrosion of metal surfaces are critically important (Galie et al., 2018).

Places Where Biofilm Forms

Infections or intoxications may arise due to food-borne diseases associated with microbial biofilms that may form on the factory equipment or food matrices. These toxins might be secreted by the biofilm that may be found within the food processing plants and contaminate food matrices and cause an individual or multiple intoxication. It has been found that human health is at risk due to the presence of biofilm in food industries.

The primary locations for the development of biofilm depend upon the type of factory and include places such as pasteurizer plates, contact surfaces, packing material, storage silos for raw materials and additives, gloves of employee, dispensing tubing, reverse osmosis membranes, animal carcasses, milk, water and other liquid pipelines, tables etc (Camargo et al., 2017; Saxena et al., 2018).

Organisms Involved in Biofilm Formation

There are various microorganisms that might grow on food industry infrastructures and food matrices and form biofilms. Many bacterial species are responsible for the colonization, initial maturation and dispersal of food industry biofilms, in addition to their health-related problems in dairy products, ready to eat foods and food matrices. These pathogens are mainly *Listeria monocytogenes* which is a ubiquitous species in water and soil which leads to abortion in pregnant women and several other serious complications in the elderly as well as in children, *Salmonella enterica* which is capable of inducing massive outbreaks and even death in elderly as well as in children when contaminating a food pipeline, *Staphylococcus aureus* which is known for their numerous enteric toxins, *Escherichia coli* includes the enterohemorrhagic as well as enterotoxigenic strains and *Bacillus cereus* that secretes the toxins and causes vomiting as well as diarrhoea symptoms. Biofilm formation may also include thermophilic *Geobacillus stearothermophilus* and psychrotrophic *Pseudomonas* sp. Due to the formation of biofilm by pathogenic species such as *Vibrio* sp., *L. monocytogenes*, *Aeromonas hydrophila*, *S. enterica* etc. fresh products of fish may suffer and causes significant economic as well as health issues (Mizan et al., 2015).

Moreover, many genomic variations have been found in the biofilm-forming microbial species with respect to key genes that involve in defining the characteristics of biofilm and at different conditions, it gives rise to different biofilms. Along with the variability of colonizing microbial species and high diversity of affected environments, this complexity complicates the eradication of biofilm in the food industry.

Mechanisms of Biofilm Formation

A precondition to formation of the biofilm is that microbes should get enough close to the surface as they approach the surface through several forces, both repulsive as well as attractive. On most of the environmental surfaces, the negative charges on the microbial surfaces face repulsion against the negative charges at distances of about 10–20 nm from the surface (Palmer et al., 2007). This repulsion might be overcome to some extent due to the attractive Van der Waal forces between the surfaces as well as the microbial cells further facilitated by the use of flagella and fimbriae that provide the necessary mechanical attachment to the surface (Palmer et al., 2007).

Attachment, maturation as well as the dispersion are the three stages that describe the formation of biofilm. This attachment step may further be categorized as a two-stage event comprising an initial reversible attachment and irreversible attachment (Renner and Weibel, 2011). According to Sutherland, (2001) this irreversibly attached biofilm tolerate stronger chemical and physical shear forces. It has been described that the flagella and type IV pili-mediated motilities are important in the initial attachment stage. Flagella are dangerous for the initial interactions between the surfaces and cells. To form microcolonies as well as aggregate, the type IV pili-mediated twitching motilities enable the attachment of cells. It has been reported that type IV pili-deficient mutants lack the ability to form microcolonies; similarly, the flagella-deficient mutants *Pseudomonas aeruginosa* could not anchor themselves onto surfaces (O'Toole and Kolter, 1998).

The initial step for the formation of biofilm for human pathogens such as *S. aureus* and *S. epidermidis* is the attachment to human matrix proteins like vitronectin (Vn), fibrinogen (Fg), fibronectin (Fn) etc. Adhesive matrix molecules have been recognized by the bacterial surface components that are covalently linked with the peptidoglycan present in cell wall. *S. epidermidis* has only 12 microbial surface components that recognize the adhesive matrix genes molecules whereas *S. aureus* has more than 20 microbial surfaces (Otto, 2008). The initial attachment of biofilms is also mediated by the non-covalent adhesions like the autolysins (Heilmann et al., 1997; Karn et al., 2020). Bacterial attachment to a surface signifies the irreversible phase for the production of extracellular polymeric substances matrix. The extracellular polymeric substance matrix of *P. aeruginosa* has been studied well and its biofilm seems to play a vital role in the progression of cystic fibrosis.

Microbes like *Pseudomonas aeruginosa* that are attached to a surface also produce a major polysaccharide component Alginate part of extracellular polymeric substances matrix in quantities that are several folds more than by planktonic cells (Davies et al., 1993). For the production of alginate, the σ factor AlgT is required, down regulates flagella genes (Garrett et al., 1999; Karn et al., 2020). Once the first layer of the biofilm gets established, the cells of similar species or other species are recruited from the bulk fluid to the biofilm. It has been found that the biofilm grows as tower shape or from a thin layer to a mushroom structure. Microbes are arranged according to their aero tolerance as well as metabolism in a thick biofilm of greater than 100 layers such as to avoid exposure to oxygen, anaerobic microbes prefer to live in a deeper layer. Within the biofilm communities' bacteria tend to communicate with each other and take specialized functions. As the biofilm gets matured, more scaffolds of biofilms like polysaccharides, proteins, DNA etc. are secreted by the entrapped microbes into the biofilm. After the maturation of biofilm, the dispersal step begins that is also critical for the life cycle of biofilm. In this step, biofilms get dispersed due to a myriad of factors like out grown population, lack of nutrients, intense competition etc. Dispersal can occur either in part or the whole of biofilm structure. Further, the release of planktonic microbes helps in the initiation of new biofilms at other sites.

Economical Loss Due to Biofilm Formation

Biofilms are the combinations of bacterial cells that spoil and contaminate industrial environments as well as their different structural units. These bacterial cells having extracellular polymeric substances colonizes the living as well as non-living surfaces and poses serious problems for all the industries causing tremendous economic loss, in terms of the quality of products as well as affecting their processes. There are many industries such as marine, medical, dairy, food, water, power plants, wine that get adversely affected due to biofilm formation. Reduction in the efficiency of heat-transfer, pipe blockages as well as water logging hampers the operating system of plants. Many industries do not set up any remedial measures to control the formation of biofilm, as most are not aware of this threat (Karn, 2011; Vishwakarma, 2019).

Chapter 3

Microbial Diversity and Their Importance

Abstract

> Microbial diversity is the group of microorganisms that are found almost everywhere and comprises smallest unicellular life forms such as bacteria, eukaryote and algae/fungi etc. Variety, variability as well as number of living organisms are defining characteristics of microbial diversity. These microorganisms play an important role in the corrosion process. In the chapter, we discussed about the microbial diversity and the details of microorganisms involved in the corrosion process such as Iron-oxidizing bacteria, Iron-reducing bacteria, Manganese-oxidizing bacteria, Manganese-reducing bacteria, Sulphur-reducing bacteria and Sulfur-oxidizing bacteria and various other microorganisms that are involved in the corrosion process by way of diverse metabolic/biochemical activities.

Keywords: microbes, diversity, bacteria, biochemical characterization

Introduction

The range of various kinds of organisms such as archaea, bacteria, fungi and protists are present in the environment. There are several different bacteria that flourish all over the biosphere and describes the limits of life and create the conditions conducive for the evolution as well as the survival of other living organisms. Due to their different physiological characteristics, cellular metabolism and morphology; various ecological activities and distributions and also by distinct genomic evolution, structure and expression, different kinds of organisms get distinguished. Currently, the microbial diversity that exist son the earth is known to be enormous and high, though it has been found that the true extent of microbial diversity is largely unknown (Dunlap, 2001; Karn et al., 2010 a, b; 2011).

Microbial Diversity

Microbial diversity denotes biological heterogeneity in natural ecosystems reflecting the pressure of variability of living organisms. Microbial diversity can be defined in terms of ecosystems, species as well as genes equivalent to their hierarchy as well as fundamental levels related to biological organization (Magurran, 2004) found globally such as on ocean surface and subsurface, tundra region, tropical forests, deserts as well as temperate forests and even in the places that were formerly considered uninhabitable by any form of life including sulfurous or thermal springs as well as the polar ice caps.

Among the vast array of microorganisms, archaea, eukaryotes and bacteria are the three primary groups of microorganisms. Archaea and Bacteria are the prokaryotes as their genetic material held in a single chromosome. Most of the genome in eukaryotes held in multiple chromosomes. Due to the microscopic identification of bacterial metabolic activity and shape of the cells, DNA sequencing, gram-staining techniques as well as by genetic identification of RNA, over 11,000 species have been found. Out of which 500 species named as archaea that can be divided into two phyla i.e., crenarchaeota as well as euryarchaeota. Eukaryotes have eight super groupings and all of them are single-celled organisms while five bear completely microbial features (Money, 2014).

Diversity is the main descriptor of community structure and is a determining factor in the dynamics and function of ecological communities (Loreau, 2010). However, the latter implicitly differs based on the different types of organisms and their relative abundance within the community which is considered to be significant (Fakruddin and Mannan, 2013).

In this planet, the presence of overall biological microbial diversity is a vital constituent and the diversity we are observing today is the consequence of around 4 billion years of evolutionary change. For the diverse bacterial communities, freshwater lakes are the vital habitats separately from several niches present in the earth. The marine as well as freshwater environment exhibit variation change in terms of depth, salinity, nutrient content as well as average temperature, nevertheless, both provide excellent habitats for microorganisms. It has been found that the environment of freshwater provides an essential environmental resource as well as diverse ecological habitats. In the freshwater sediments, prokaryotes are amongst vital contributors to the transformation of minerals as well as complex organic compounds (Jurgens et al., 2000; Nealson, 1997).

Microorganisms Involved in the Corrosion Process

In the corrosion process, microorganisms play a vital role and involve many genera and species. Microorganisms can be divided into three groups such as algae, bacteria and fungi. Bacteria that involve in the sulfur cycle play a vital role in corrosion process especially those that involve in the redox reactions of Sulphur. It has been found that sulphate reducing bacteria is the most important bacteria found in microbial corrosion processes. Other organisms such as Iron-oxidizing, Iron-reducing, Manganese-oxidizing, Manganese-reducing (Myers and Kenneth, 1988), *Bacillus* sp. (Karn, 2017) and many other genera are involved in the corrosion process. A fungus such as *Cladosporium resinae,* finds special mention in the corrosion of aluminum from the fuel tanks of subsonic aircrafts as well as wing perforation. The corrosion of aluminum is mainly caused by the production of carboxylic acid. Algae appear to have the potential for inducing corrosion by virtue of their role in production of oxygen, corrosive organic acids and nutrients for other corrosive microorganisms (Iverson, 1987). Figure 3 represents the microbes involved in the corrosion process.

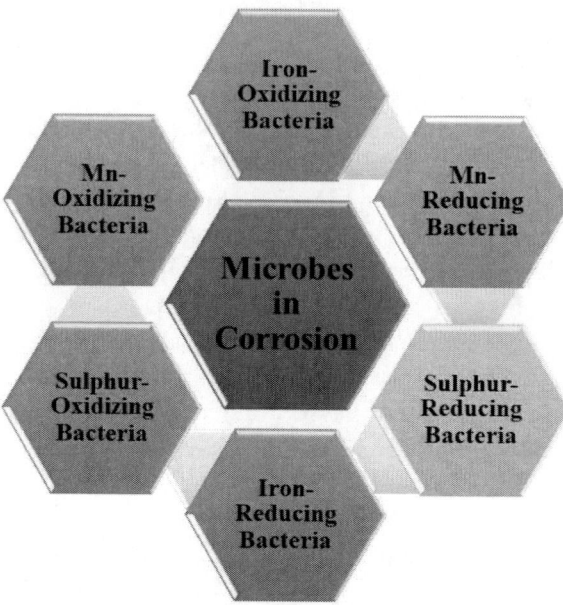

Figure 3. Microbes involved in the corrosion process.

Iron-Oxidizing Bacteria

In order to obtain metabolic energy, iron-oxidizing bacteria namely *Leptothrix* sp., *Sphaerotilus* sp., *Gallionella* sp. and *Crenothrix* sp. Oxidizes ferrous ions into ferric state and deposit the ferric oxide on the carbon steel pipeline surfaces as well as promote 'tubercle' formation. In the long sheaths, the most common iron-oxidizing bacteria have been found which belongs to the class Chlamydobacteriales. These long filaments have a characteristic pattern and can be observed readily under the microscope. In iron and carbon steel distribution system pipelines, these filamentous iron bacteria are omnipresent and are generally reported in the deposits linked with tuberculation.

One of the most typical characteristics of *Sphaerotilus-Leptothrix* group of bacteria is the oxidation of ferrous ions. It was reported that autotrophic organisms such as *Leptothrix* sp. receive energy by oxidation of ferrous ion. *Leptothrix* sp. are gram-negative unsheathed bacterium commonly growing in iron containing water and store iron in their cellular mass. The growth behavior of *Leptothrix* sp. is similar to algae and produce high biomass, further play actively in biogeochemical process of ecological importance (Karn et al., 2017).

Just like algae, the species of *Leptothrix* is also essential for the biogeochemical as well as ecological homeostasis.

The sheath forming characteristics of these organisms indicate the ir-relationship with cyanobacteria, which under favorable conditions assist in the formation of gelatinous or fibrous sheath. Due to the impregnation and deposition of iron oxide as well as excretion of organic matter, these thin sheaths get heavier making them extremely resistant and heavy for decomposition. For the microorganisms, the presence of the sheath has nutritional as well as ecological importance and this sheath permits the microbial attachment on the solid surfaces of pipelines and due to this feature, the growth of bacteria favored in the pipelines as well as in running water (Karn et al., 2010; 2017).

Sheathed bacteria such as *Leptothrix lopholea* and *Sphaeroti lusnatans* produce a fixative agent which help them attach on the solid surfaces but not all members of iron bacteria do so. Fixing agents originates from swarmer cells when contacting a surface and then attach to the flagellum. The sheaths of *Leptothrix* sp., assist in the formation of a membrane that is relatively impermeable to oxygen, and in the process decreases the availability of oxygen thus establishing a micro-aerophillic cell. By increasing the thickness of biofilms inside the film, the surface become more anaerobic and this difference in the potential between the inner and outer surfaces accelerates the process of corrosion (Karn et al., 2020).

A broad range of microorganism's are involved in iron-Fe (II) oxidation in order to receive the reducing equivalent for the generation of PMF and NAD^+ reduction. Phototrophic organisms such as acidophilic aerobes (*Leptospirillum* sp. and *Acidithiobacillus* sp.), nitrate respiring anaerobes (*Dechloromonas* sp. strain UWNR4), (*Rhodobacter* sp.) and neutrophilic aerobes (*Gallionella* sp. and *Sideroxydans* sp.) also involved in the oxidation process (Dong et al., 2017). Anaerobic as well as phototrophic iron oxidizers are heterotrophs that uses electrons as a supplement from Fe (II) whereas aerobic iron-oxidizing bacteria are autotrophic that require electrons for carbon fixation and Fe (II) as the only source of energy.

Acidithiobacillus ferrooxidans is one of the utmost important iron-oxidizing bacteria and is known well for its impact on the environment as it produces energy by linking the oxidation of Fe (II) to the reduction of O_2 (Gong et al., 2018). Oxidation of the Fe (II)S mineral is the well-known form and known as pyrite and plays a vital role in acid mine drainage which is also the cause of acidic rivers and lakes as it contains iron in excess.

A. ferrooxidans oxidizes Fe (II) to Fe (III) with bacteria further using the liberated electrons to reduce O_2 to form H_2O. Next, resulting Fe (III) spontaneously reacts with FeS_2 (pyrite) to form excess of protons, Fe (III) and thiosulphate. The presence of sulphur-respiring bacteria, along with oxidation of thiosulphate, forms sulphate and has acceleratory effect of low pH, which in turn enhances the availability of soluble reduced iron and accelerates the pyrite dissolution (Gong et al., 2018).

Iron-Reducing Bacteria

Iron-reducing bacteria (IRB) play an important role in the transformations of biogeochemical cycles which harness environmental bioremediation processes. For the dissimilatory growth, a sub section of IRB involves direct contact with Fe (III)-bearing minerals and thus far these microbes must move amongst the mineral particles. Additionally, during bio-stimulation experiments, they proliferate in planktonic consortia (Luef et al., 2013).

Due to strictly facultative or anaerobic nature of iron-reducing bacteria, Fe (III) minerals may be decreased microbiologically in the natural systems and uses H_2 or a wide range of organic compounds as electron donors

(Ehrlich and Newman, 2009). Bacteria and Archaea such as *Shewanella putrefaciens* (strains 200 and ATCC 8071), *Geobacter sulfurreducens* (strains KN400 and PCA), *Geobacter metallireducens* (strain GS-15) and *Shewanell aoneidensis* (MR-1) have the ability to use Fe (III) as a terminal electron acceptor across the domains (Mahadevan et al., 2001; Weber et al., 2006; Fredrickson et al., 2008). Geobacteraceae, which oxidize acetate to CO_2 with Fe (III) as the sole electron acceptor (Lonergan et al., 1996) as well as the H_2-oxidizing Fe (III) reducers like *Shewanella putrefaciens* and *Shewanella alga* observes that the magnetite and Fe sheet of silicates are resistant to leaching treatment in an inorganic acid as well as commonly classified as poorly reactive or unreactive iron phases in sequential leaching schemes.

Manganese-Oxidizing Bacteria

Manganese oxides such as oxyhydroxides and hydroxides are ubiquitous in sediments as well as soil and play a vital role in the biogeochemical cycles of organic carbon as well as metals while significantly influencing the transport and fate of both the nutrients and contaminants present in the environments via oxidative, catalytic and sorptive processes (Tebo et al., 2004).

Manganese-Reducing Bacteria

For reduction of manganese, bacteria couple their growth with the metabolic activity and play vital roles in the biogeochemistry of certain anaerobic environments. For instance, *Alteromonas putrefaciens* MR-1 couples its growth only under the anaerobic conditions for manganese oxide reduction. The reduction characteristics are reliable with biological moreover not a manganese reduction, indirect chemical that suggested that this bacterium uses manganic oxides as a terminal electron acceptor. Large number of other compounds has also been utilized as terminal electron acceptor and this flexibility provides advantage in the environments where the concentration of electron-acceptor differs (Myers and Kenneth, 1988).

Sulphur-Reducing or Oxidizing Bacteria

Sulphur-reducing bacteria (SRB) and iron-reducing bacteria are also capable of reducing sulphur. *Desulfuromonas acetoxidans* is capable of reducing sulfur at the expense of acetate. SRB can also use organic disulfide molecules like glutathione or cytokine. Though sulfur and sulfate reducers can coexist, the latter can produce more sulfide. Many of these bacteria are able to generate ATP during sulfur reduction. SRB usually belongs to archaea, thermophilic and methanogenic group. Archea reduce sulphur and generates methyl. Some sulphur reducers also belong to proteobacteria. These organisms have high ecological competitiveness and metabolic flexibility (Bharathi, 2008).

Chapter 4

Techniques to Determine the Microbial Diversity

Abstract

Diversity of microorganisms is essential to maintain balance of nature. Therefore, various techniques are being developed to study the diversity of organisms. Microorganisms can be characterized based on their evolution, ecological distributions and activities, distinct genomic structures and expressions. Identification of microorganisms done on the basis of their morphological characteristics, biochemical characterization and molecular characterization such as random amplification of polymorphic DNA, Ribosomal intergenic spacer analysis/automated intergenic spacer analysis, PCR-independent approaches, REP-PCR (BOX Element), DNA Reassociation kinetics and DNA:DNA hybridization, Reverse sample genome probing, DNA microarrays, next-generation sequencing, matrix-assisted laser desorption/ionization time-of-flight mass spectrometry (MALDI-TOF), Flow cytometry method. By one estimate, 99% of the organisms on this planet remain unknown due to lack of suitable medium for their culture, nutrient and environmental conditions. Recently, metagenomic approaches has emerged as an important modern tool to overcome the limitations and can be considered as a culture independent process.

Keywords: biodiversity, bacteria, fungi, PCR, RNA, MALDI-TOF, RAPD, REP-PCR, ARISA, RISA, SSCP, hybridization, RSGP, gene sequencing, flow cytometry

Introduction

Microbial diversity comprises a range of various unicellular organisms such as fungi, bacteria, protists and archaea. There are diverse microbes that thrive under different environments which defines the limits of life. This also creates the conditions conducive for evolution as well as survival of other organisms. Microorganisms are distinguished by their characteristics on the basis of their morphology, physiology, cellular metabolism and by their distinct genomic structure, evolution and expression and also by several ecological distributions and activities.

Enormous microbial diversity is present on the earth, nevertheless, the true extent of diversity is unknown. To some extent, microbial diversity can be explored as permitted by new molecular tools which rapidly analyze their characteristics and evolutionary relationships (Dunlap, 2001).

Morphological Characterization of Bacteria

It is a well-known fact among microbiologists that only a minority of microorganisms are culturable and characterized. It is difficult to classify the prokaryotic organisms and the validity of the classification has been questioned frequently. Morphological characterization of microorganisms has been based on Gram staining, cell shape, movement, cell wall, flagella etc. which may not be sufficient for creating a detailed classification/profile of microbes (Giovannoni et al., 1990). This is the reason why it is essential to look for some other adequate techniques for further classification.

Biochemical Characterization

Microorganisms, particularly human pathogens, often, can be characterized or identified by some unique carbohydrates present on their cell wall or plasma membrane. Antibodies and other carbohydrate-binding glycoproteins present in the host system can attach to these specific carbohydrates and cause the cells to clump together. Thus, serological tests have been developed such as Lancefield groups tests for the identification of *Streptococcus* species. Biochemically, biodiversity may also be determined by gram staining technique, indole production, catalase test, Voges Proskauer test, sugar fermentation, urease production, citrate utilization test, oxidase test, phenylalanine deaminase test, methyl red test etc.

Fatty Acid Methyl Ester (FAME) Analysis

To assess the structure of microbial community, phospholipid fatty acid (PLFA) analysis has been used as a culture-independent method.

Determination of the phospholipid fatty acid (PLFA) profiles provides for a broad diversity measurement of microbial community at the phenotypic level (Chayani et al., 2001). On the basis of grouping of fatty acid present on the microbial surface, this method provides information on the bacterial community composition. Fatty acids are used as a chemotaxonomic marker and is regarded as a signature molecule found in all living cells. Phospholipids are the key determinant found in the cell membrane of microorganisms. Fatty acids are also an important component within the microbial community which can help differentiate the changes in the microbial population in an environmental sample (Eiland et al., 2001).

Molecular Methods to Determine Microbial Diversity

Molecular-Based Approach for Analyzing Microbial Diversity

Traditionally the culture techniques have been in practice to yield information about the microbial world, though, only a limited fraction of all microorganisms existent for the isolation and enrichment of microbes. Polymerase chain reaction-based molecular methods are fast and sensitive alternatives to conventional methods. Molecular methods seize an opportunity to analyze the community of microbes in its full diversity on the basis of the analysis of single cell (Figure 4).

Genetic fingerprinting techniques are used in the study of population structures and dynamics (Fakruddin et al., 2013). Different molecular approaches have been established by taxonomists to understand the community as well as diversity composition of natural environment microflora which permits quick profiling of bacterial communities to provide the information about the presence of specific phylogenetic groups. PCR-based fingerprinting methods of bacterial communities consists of the extraction of nucleic acids, amplification of rRNA/rDNA and also the PCR products analysis by fingerprinting techniques.

Initially, the PCR-based approaches such as molecular approaches relied on the target genes colony from the environmental samples for the ecological studies (De Santis et al., 2007). The information of prokaryotic diversity provides the PCR-based 16S rDNA profile that allows prokaryotes identification and prediction of phylogenetic relationships (Pace, 1997, 1999). Consequently, the detailed information about the community structure of an ecosystem in terms of composition, richness as well as evenness has been provided by 16S rDNA-based PCR techniques like amplified ribosomal DNA restriction analysis, ribosomal intergenic spacer analysis, temperature gradient gel electrophoresis, denaturing gradient gel electrophoresis, terminal restriction fragment length polymorphisms single-strand conformation. These techniques can also be used in the comparison of diverse species present in a sample (Rawat and Johri, 2014).

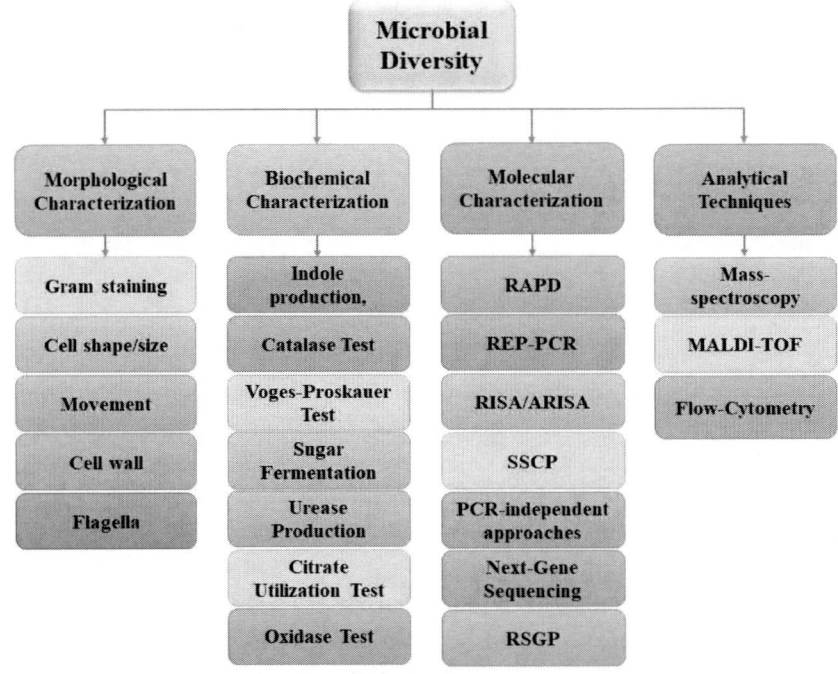

Figure 4. Techniques involved in microbial diversity.

Amplified Ribosomal Restriction DNA Analysis (ARDRA)

Amplified ribosomal DNA restriction analysis is based on variations of DNA sequences that is present in PCR-amplified 16S rRNA genes (Smit et al., 1997). The products of PCR which amplified from environmental DNA is commonly digested with tetra cutter restriction endonucleases such as HaeIII and AluI as well as also resolved the restricted fragments on polyacrylamide gels or agarose. Even though amplified ribosomal DNA restriction analysis delivers slightly or no information about the presence of which type of microorganisms in the sample, over the time this method is still suitable for rapid monitoring of microbial communities, or to compare the diversity of microbes in response to change in the environmental conditions. To estimate OTU (Operational Taxonomic Unit) in environmental clone libraries and also to identify the unique clones, amplified ribosomal DNA restriction analysis is used on the basis of restriction profiles of clones (Smit et al., 1997).

Amplified ribosomal DNA restriction analysis (ARDRA) is used to study the microbial diversity which depend on DNA polymorphism. On the basis of these repetitive units of nuclear ribosomal DNA (rDNA), this technique of amplified ribosomal DNA restriction analysis consists of conserved coding and variable non-coding regions and these regions amplified by PCR, restriction fragments are separated according to their sizes by using gel electrophoresis and the amplicon is digested by restriction endonucleases. Liu et al., (1997) described that the PCR-amplified 16S rDNA is digested with a 4-base pair cutting restriction enzyme whereas according to Pace, (1999) to screen the clones and also to measure the microbial community structures, banding patterns of gel electrophoresis can be used. It has also been studied that to detect the structural changes in the bacterial community this is useful nevertheless not as an amount of diversity or detection analysis is a sensitive technique and gives high resolution to deliver consistent genotypic of specific phylogenetic groups (Liu et al., 1997).

Amplified ribosomal DNA restriction characterization at the community level of compost microbes (Heyndrickx et al., 1996). To recognize the community structure from various samples like microbes that present in self-heating phase of composting materials (Koschinsky et al., 1999) as well as in casing (Choudhary et al., 2009; Choudhary, 2011) and feathers (Tiquia et al., 2005), this technique has been used frequently. The universal primers ITS 1 and ITS 4 are uses by ARDRA-ITS (also termed ITS-RFLP) that anneal to the evolutionary stable 18S and 28S rRNA genes (White et al., 1990). The fungi investigation has been allowed by their attachment with conserved rDNA regions without the previous knowledge of their genome organization. Due to non-coding variable internal transcribe spacers ITS I (between 18S and 5.8S) and ITS II (between 5.8S and 28S), the conserved domains get interrupted that provides information for the differentiation.

At the species as well as subspecies levels, the ARDRA-ITS exhibits their differences. It has been stated that the studies on indoor basidiomycetes are rare. In T-RFLP, firstly, the extraction of DNA takes place from the bacterial communities, and then uses it as a template for PCR amplification of desired gene(s). For the microbial composting, the 16S rDNA is used routinely with the appropriate primers. According to Choudhary et al., (2009) For the estimation of microbial community structure, the amplified DNA is digested with restriction enzymes as well as the terminal fragment size is generated by each amplicon that is used without the need of sequencing the terminal fragments. For the estimation of genetic diversity of microbial community, the abundance as well as the size distribution of DNA fragments has been characterized. Nevertheless, for the estimation of phylogenetic positions of community members, this method is not pertinent. The analysis of T-RFLP of bacterial community reveals the extensive microbial diversity. For the identification of casing soil bacteria, molecular tools have been used as well as 16S rRNA gene analysis used intensively to understand the phylogenetic relationships (Choudhary et al., 2009).

Random Amplification of Polymorphic DNA (RAPD)

On the basis of polymerase chain reaction (PCR), the investigation of randomly amplified polymorphic DNA (RAPD) uses short (about 10 bases) randomly chosen single primers that anneal as reverted repeats in the genome to the complementary sites (Agrawal and Shrivastava, 2013). PCR amplifies the DNA between the two opposite sites by the primers at starting as well as at end points. Further, the amplification products get separated on gels as well as due to the presence or absence of bands (polymorphism) the banding patterns separate the organisms. It also gets discriminates at different taxonomical levels such as the primer used, investigated fungus as well as species and isolates by RAPD analysis.

According to Williams et al., (1990) this random amplification of polymorphic DNA (RAPD) method is used to make a single random primer at low stringency for polymorphic DNA amplification. At various sites, the primer at low stringency anneals to the target DNA whose sequences cannot be exactly complementary to the primer sequence. Numerous distinct bands of DNA get amplify upon the annealing of primer in inverted orientation at distances appropriate for the amplification. Even though the analysis of RAPD is convenient as well as rapid but not reproducible and even it can change the fingerprint by small variations in the batch of buffer or Taq polymerase. Due to RAPD technique, the conditions for the direct DNA amplification enhanced case-by-case in the natural environment. Singh et al.,

(2005) carried out that the analysis of ITS region of 5.8S rRNA gene from eight *H. grisea* isolates by RAPD technique. In this region, intra-specific diversity gets visualize. Within these species, the isolates exhibit the genetic differences that correlate with morphological variation.

REP-PCR (BOX Element)

Repetitive extragenic palindromic-PCR (REP-PCR) technique is used to obtain the genomic DNA fingerprint of bacteria and for this, they use the primers to match the short consensus repetitive sequences. According to Gomez et al., (2000) there are three different primers that are used such as REP (*Escherichia coli*), BOX (*Streptococcus pneumoniae*) and ERIC (*Salmonella typhimurium*).

Polymorphism has been represented by the differences in the band sizes in the distances between the repetitive elements of unlike strains. Versalovic et al., (1991) describes that the repetitive extragenic palindromic-PCR (REP-PCR) is a genotypic process which uses the oligonucleotide primers that are complementary to repetitive sequences spread all over the genome of *E. coli*. Diverse regions of DNA amplify by using the PCR and flanked up by rep sequences which leads to amplicon patterns that are specific for an individual *E. coli* strain (Rademaker and de Bruijn, 1997).

According to Versalovic et al., (1991) the conserved repetitive sequences are divided into four types such as enterobacterial repetitive intergenic consensus (ERIC) sequences, poly-trinucleotide (GTG) 5 sequences, repetitive extragenic palindromic (REP) sequences and BOX sequences. There are five REP-PCR methods that are frequently used for genotyping of various microbial strains namely ERIC2-PCR (primer ERIC2), REP-PCR (primer sets Rep1R-I and Rep2-I), (GTG)5-PCR [primer (GTG)5], BOX-PCR (primer BOX A1R) and ERIC-PCR (primer sets ERIC1R and ERIC2). Among the isolates, higher degree of resolution has been produced by the BOX-PCR which is the multilocus technique whereas the repetitive sequences that are located erratically inside the whole genome in the form of BOX elements and BOX primers amplify the genomic regions among the two BOX elements. The distribution of these repetitive sequences (BOX and ERIC) is the true reflection of genomic structure as well as the amplification of inter REP elements frequently detects the comparisons in given group of microbes (Selenska-Pobell et al., 1995).

Versalovic et al., (1991) found that ERIC and REP sequences are almost ubiquitous in bacteria and helps in rapid molecular characterization by PCR-based fingerprinting. Choudhary et al., (2009) describes that REP-PCR used for the bacterial identifications it provides genomic fingerprint of chromosome structure that considered variables between the strains. Both eukaryotic as well as prokaryotic genomes comprises discrete repetitive sequences which ranges from 15 to several hundred base pairs in length. As compares to longer repeated elements that contains coding sequences, these elements are non-coding nevertheless they are present in high copy. In bacteria, the interspersed repetitive sequences that describes are ERIC sequences (124 bp), Rep sequences (386 bp) and BOX elements (154 bp) and these sequences can be diagnostic and allows the differentiation down to the strain or species level. All the isolates of microbes get recovered from diverse stages of sample subjected to REP-PCR analysis by BOX primer. A distinctive banding pattern detected in the BOX element. On the basis of UPGMA, the dendrogram of banding patterns shows the substantial discriminatory relationship amongst the isolates.

Ribosomal Intergenic Spacer Analysis (RISA)/ Automated Intergenic Spacer Analysis (ARISA)

RISA and ARISA provide ribosomal-based fingerprinting of the microbial community. In RISA and ARISA, PCR amplifies the intergenic spacer (IGS) region between 16S and 23S ribosomal subunits, denatured and then separated on polyacrylamide gel under the denaturation conditions. tRNA has been encoded on IGS region and is useful to differentiate between the microbial strains and due to the heterogeneity of IGS length and sequence, it is also closely related to the species (Ranjard et al., 2000; Fisher and Triplett, 1999). With each band, RISA delivers a community-specific profile to one organism in the original community.

ARISA is an automated version of RISA which involves the use of ISR fragments as well as fluorescence-labeled forward primer that gets detected automatically by a laser detector. Many samples can be analyzed simultaneously by ARISA and can be exploited to exhibit the bacterial diversity as well as richness (Fisher and Triplett, 1999). Amongst the eubacteria, the 16S rRNA sequences are very conserved (Woese, 1987) and in this region, the analysis of genetic variation is not suitable to discriminate the strains within the species. In the compost ecosystem, the ribosomal intergenic sequence analysis (RISA) is used to analyze the composition of species (Saison et al., 2005). In the initial stage of composting, there is a total change in the fungal community structure which is characterized by employing F-ARISA and 18S rRNA sequencing and cloning (Hansgate et al., 2004). The analysis of polymorphism of intergenic spacer length between rrs (16S rRNA) and rrl (23S rRNA) genes involved in this technique whose sizes differ from 50 bp to more than 1.5 kb depends upon the species. The subsequent amplicon sequencing allows the taxonomic identification of specific populations inside a community.

On the basis of electrophoresis of PCR-amplified 16S rDNA fragments, denaturing gradient gel electrophoresis (DGGE) or temperature gradient gel electrophoresis (TGGE) examines the genetics of microbial diversity (Muyzeret al., 1993). In DNA sequences, the point mutation has been detected by DGGE and TGGE. DNA is extracted from the sample and is further amplified by using PCR having universal primers with targeting against 16S or 18S rRNA sequences. 35-40 base pair GC clamp at the 5'-end of forward primer is meant to confirm that at least few parts of DNA remain double stranded. On the basis of melting behavior of double-stranded DNA, the separation arise son a polyacrylamide gel with a gradient of increased concentration of denaturants i.e., urea and formamide DNA melts upon denaturation that are specific for sequences as well as differentially it migrates through the polyacrylamide gel.

Molecules having diverse sequences can differ in their melting behavior; consequently, they stop at different positions in the gel (Muyzer and Smalla, 1998). Further, SYBR Green I, silver staining or ethidium bromide has been practiced visualizing the bands of DNA in DGGE or TGGE profiles being more sensitive than others. Nevertheless, it also stains single-stranded DNA, though, it can get digested by nuclease to diminish the interference. In environmental microbiology, as compared to DNA extraction methods, DGGE or TGGE has been commonly used to study the community complexity, analysis of enrichment cultures, isolate bacteria, monitor population shifts and to detect the sequence heterogeneity of 16S rRNA genes as well as 18S rRNA (Muyzer and Smalla, 1998; Nicolaisen and Ramsing, 2002).

DGGE technique is also used in the detection of non-RFLP polymorphism (Bodelier et al., 2000). The variations in sequences have been utilized by this technique in PCR-amplified DNA fragments of identical length and determine them on the basis of variances in their mobility in polyacrylamide gels which contain the gradient of denaturing agent (Muyzer et al., 1993). According to their molecular weight, fragments start to move initially but as they reach higher denaturing conditions each reaches a point where DNA starts melting.

A valuable approach has to be obtained by 16S rRNA gene profiles for the identification of temporal and spatial differences in microbial community structure in the compost ecosystem (Takaku et al., 2006; Ishii et al., 2000). By using phylogenetic probes or sequencing, the rRNA genes get combined with hybridization through DGGE analyses technique to obtain the assessment of the phylogenetic affiliation of numerically dominating members of community. During the laboratory scale composting process of garbage, DGGE amplifies the fragments of 16S rDNA that have been used to analyze the bacterial succession (Ishii et al., 2000). Separation of PCR as well as RT-PCR products by DGGE analysis and identify by hybridization with hierarchical set of oligonucleotide probes that are designed to detect the ammonia oxidizer like sequence cluster in genera *Nitrosomonas* and *Nitrospira* (Kowalchuk et al., 1999). In the composting process of rice straw, the phylogenetic as well as the succession profile of eukaryotic communities' studies using DGGE followed by 18S rDNA (Cahyani et al., 2003).

TGGE uses a uniform concentration of denaturant in the gel instead by using a gradient of the denaturant as well as uniformly enhances the temperature with time during the electrophoresis. This results in more easily reproducible separations than those usually obtained with DGGE. TGGE detected most active microbe by the rRNA amplicon which are obtained by RT-PCR (Felskeet al., 1998). By using DGGE and TGGE (Bruns et al., 1999) determined the temporal as well as spatial diversity of ammonia oxidizers in succession, native and tilled soils. In the comparison of DGGE and TGGE with T-RFLP, each amplicon isolated and characterized within a microbial community profile by DNA sequencing. Consequently, samples can be compared not only by their profile patterns but selected components can also be clearly identified. But DGGE can only determine the microbes that constitute up to 1% of the total bacterial community (Zoetendal et al., 2004). Therefore, the separation of amplicons by DGGE may not be perfect and amplifying-based sequence analyses need careful interpretation (Nikolausz et al., 2004).

Single-Strand Conformation Polymorphisms (SSCP)

Single-strand conformation polymorphism (SSCP) analysis is a method that is used to detect the sequence differences of single-stranded DNA (ssDNA) by non-denaturing polyacrylamide gel electrophoresis (PAGE). Generally, PCR amplification, sample resolution by non-denaturing PAGE as well as denaturation of the double-stranded PCR product with formamide and heat (or other denaturants) has been involved in the SSCP process of the target DNA (Orita et al., 1989). On the basis of their primary sequences, the ssDNA fragments are expected to fold into a three-dimensional shape during the electrophoresis.

There are various reports that have suggested that there might be difference of only a single nucleotide in the sequence between the wild-type sample and a mutated fragment and a unique and distinct electrophoretic mobility pattern is adopted by each sequence. Consequently, non-denaturing PAGE is used to separate the same size of complex mixtures of DNA species of the same size into bands of different mobilities, because of the difference in their predominant semi-stable conformations. In DNA the point mutations as well as novel polymorphisms are detected by SSCP-PCR (Orita et al., 1989). This technique is used to distinguish the same size of DNA molecules having different sequences of nucleotide by using gel electrophoresis in non-denaturing polyacrylamide gels because of the differences in mobility that are caused by their folded secondary structure.

SSCP also distinguishes the PCR product of same sizes with different base sequences that makes the process a promising tool for the analysis of compost bacterial community at the ribosomal gene level (Rawat et al., 2005). In the compost, the diversity of fungal and microbial communities can be analyzed by using single-strand conformation polymorphisms (SSCP) of approximately 400 bp PCR products that gets amplified with the universal primer for 18S rRNA (fungi) as well as 16S rRNA (bacteria) having compost DNA as a template (Peters et al., 2000). This technique is an alternative and possibly an improvement over DGGE and TGGE. The construction of gradient gels does not require by SSCP thus enhanced the reproducibility of gels. While specific equipment is needed for TGGE and a temperature gradient incubation system for electrophoretic gels. Regular electrophoretic chambers with temperature control can be used for SSCP. As compared to DGGE or TGGE, another advantage of SSCP is that compatible primers for SSCP are easier to design, since consideration regarding the primer bias for the formation of a GC clamp during the PCR process is not required (Droffner and Brinton, 1995).

PCR-Independent Approaches

The technique generates a fractionated profile of the entire community DNA and indicates relative abundance of DNA (hence taxa) as a function of G - C content. The total community DNA is physically separated into highly purified fractions, each representing a different G - C content that can be analyzed by additional molecular techniques such as DGGE/ARDRA to better assess total community diversity. It provides a coarse level of resolution. As different taxonomic groups may share the same G - C content, its analysis is not influenced by PCR biases and since it includes all extracted DNA and uncovers even rare members in a microbial population.

DNA-Reassociation Kinetics and DNA: DNA Hybridization

Whole-genome DNA-DNA hybridization (DDH) offers true genome-wide comparison between organisms. A value of 70% DDH was proposed as a recommended standard for bacterial species delineation (Goris et al., 2007). Typically, bacterial species having 70% or greater genomic DNA similarities usually have 97% 16S rRNA gene sequence identity. Although DDH techniques have been originally developed for pure culture comparisons, they have been modified for use in whole microbial community analysis.

In the DDH technique, total community DNA extracted from an environmental sample is denatured and then incubated under conditions that allow them to hybridize or re-associate. Nucleic acid hybridization is a process wherein two DNA or RNA single chains (mono-stranded) from different biological sources form a double catenary configuration, based on contingent sequence homology between two sources, resulting in DNA-DNA, RNA-RNA or DNA-RNA hybrids. The purpose being identification or localization of certain nucleic acid sequences in the genome of some species. Two basic notions are used: the target molecule representing the DNA, RNA or protein sequence that has to be identified and the probe molecule that identifies the target, by hybridization.

When hybridization takes place on a solid substrate, it is blotting and can be divided into three categories: Southern blotting where DNA molecules are identified using DNA or RNA probes; Northern blotting where RNA molecules are identified using RNA or DNA probes; Western blotting where protein sequences are identified using specific antibodies. DNA reassociation is used to measure genetic complexity of the microbial community and has been applied to evaluate environmental diversity. Total DNA is extracted from the environmental sample, purified, denatured and allowed to reanneal.

The rate of hybridization or reassociation depends on the similarity between DNA sequences. (Theron and Cloete, 2000) have reported that as the complexity of diversity in DNA sequences increases, the rates at which DNA re-associates decrease. Similarity between communities of two different samples can be studied by measuring the degree of similarity between DNA through hybridization kinetics (Griffiths et al., 1999). Nucleic acid hybridization using specific probes is an important qualitative and quantitative tool in molecular bacterial ecology (Clegg et al., 2000; Theron and Cloete, 2000). This approach can be applied on extracted DNA or RNA under in situ conditions. Oligonucleotide probes designed from known sequences ranging in specificity from domain to species can be tagged with fluorescent markers at the 5' end (Theron and Cloete, 2000). However, dot blot hybridization is used to measure the relative abundance of a certain group of microorganisms. It provides valuable information regarding spatial distribution in microbial community or environmental samples.

Broad-scale analysis of community DNA, using techniques such as DNA-reassociation kinetics, provides information about the total genetic diversity in compost microbial community (Torsvik et al., 2002). This approach is based on the assumption that more complex denatured DNA re-associates at a slower rate than less complex denatured DNA, and that the kinetics of reassociation is proportional to the genomic complexity. The advantage of this approach is that it may be the only means developed to date by which total number of bacterial species in the sample can be estimated. Requirement of large quantity of DNA in this technique could be the major limitation because it is often technically difficult to obtain high DNA yield from soil sample. DNA: DNA hybridization provides a reasonably good means of comparing two microbial communities, although it may suffer from the lack of sensitivity that makes quantitative comparison of communities possessing similar structures difficult. Similarly, shift in GC content can be used to detect changes in bacterial community but does not provide any information regarding richness (number of species), evenness (relative abundance) and composition of the microbial community (Torsvik et al., 2002).

DNA Microarrays

DNA microarrays have been used primarily to provide a high-throughput and comprehensive view of microbial communities in environmental samples. The PCR products amplified from total environmental DNA are directly hybridized to known molecular probes, which are attached to the microarray substrate (Gentry et al., 2006). After the fluorescently labeled PCR amplicons are hybridized to the probes, positive signals are scored by the use of confocal laser scanning microscopy. The microarray technique allows samples to be rapidly evaluated with replication, which is a significant advantage in microbial community analyses.

In general, the hybridization signal intensity on microarrays is directly proportional to the abundance of the target organism. The latest development in the technology have been application of DNA microarrays to detect and identify bacterial species or to assess microbial diversity (Cho and Tiedje, 2001; Green and Voordouw, 2003). This rapidly characterizes the composition and functions of microbial communities because a single array may contain thousands of DNA sequences with the possibility of very broad hybridization and identification capacity. The microarrays can contain specific target genes such as nitrate reductase, nitrogenase or naphthalene dioxygenase to provide functional diversity information or can contain (DNA fragments with less than 70% hybridization) representing different species found in the environmental sample (Green and Voordouw, 2003).

Reverse Sample Genome Probing (RSGP)

RSGP is used to analyze microbial community composition of the most dominant culturable species. RSGP includes four steps: (1) isolation of genomic DNA from pure cultures (2) cross-hybridization testing to obtain DNA fragments with less than 70% cross-hybridization (3) preparation of genome arrays on a solid support and (4) random labeling of a defined mixture of total community DNA and internal standard. RSGP is a useful approach when diversity is low, but several molecular biologists face difficulty while assessing community composition of diverse habitats (Green and Voordouw, 2003).

Phylogenetic Analysis and 16S rDNA Gene Sequencing

Taxonomy based on comparative phylogenetic analysis of 16S rRNA genes, first introduced by Carl Woese, presents a radical departure from classical taxonomy. Cellular life forms could be divided into three primary phylogenetic domains: Archaea, Bacteria and Eucarya (Woese et al., 1990; Olsen et al., 1986). For microorganisms, molecular data often provide considerable information because microorganisms such as bacteria simply do not have the diversity of form to make morphological characteristics useful in establishing phylogeny. Aside from derivation of taxonomies, molecular phylogenetic analyses are important in identifying similarity between organisms, providing an ability to understand physiology and ecology of non-culturable species. PCR-based 16S rDNA profiling provides information about microbial diversity and allows identification of microbes and prediction of phylogenetic relationships (Pace, 1997; Song et al., 2001). Initially, molecular approaches for ecological studies relied on cloning of target genes isolated from environmental samples (Muyzer and Smalla, 1998). Therefore, 16S rDNA-based PCR techniques such as DGGE, TGGE, SSCP, ARDRA, T-RFLP, RISA and others can provide detailed information about community structure of various ecosystems in terms of diversity, invariability and constitution.

Nucleic acid sequencing provides larger discrimination than other methods and better characterization of a particular member of community (McCaig et al., 2001). 16S rRNA sequences have been used most extensively to examine the biodiversity and to construct a phylogenetic tree (Swofford et al., 1996). The phylogenetic approach for the systematic assessment of culturable microbial diversity up to the taxonomic level using nucleic acid hybridization and 16S rDNA sequences analysis has been of immense utility in the phylogenetic redraw of the classification of prokaryotic organisms (Woese, 1987). The total DNA or rRNA extraction from compost sample followed by 16S (Prokaryotes) and 18S (eukaryote) analyses does not however, always reflect the presence of qualitative and quantitative diversity (Gurtner et al., 2000; Ishii et al., 2000).

Next-Generation Sequencing

Next-generation sequencing (NGS), otherwise high-throughput sequencing, resulted in a breakthrough in the automation and commercialization of the sequencing process. In 2000, the company Lynx Therapeutics launched the first fully automated sequencing apparatus, the principle of which was still based on the Sanger method. In 2004, the company 454 Life Sciences has developed and successfully launched the sale of second-generation sequencer, which used pyrosequencing method discovered in 1996. In addition to the huge success brought about by the technique, the cost of sequencing decreased six-fold in comparison with the device from 2000s (Schuster, 2008; Brenner et al., 2000). High-throughput sequencing is probably the fastest growing method used in biotechnology. A series of modifications have been implemented which resulted in the development of relatively cheap and efficient equipment set-up. A large selection of sequencing systems has been introduced by many companies in the market, but the Illumina sequencing system finds special mention. It is the most common method in the study of metagenomes from different environments. The technology of NGS has undergone very dynamic developments over the years and performance data and bandwidth have become outdated several times.

DNA prepared for sequencing must meet several requirements. First of all, it must be free from contamination and PCR inhibitors such as humic acids, ethanol, and phenol compounds. A very important and crucial step in the preparation of biological samples is appropriate DNA extraction and its purification. Commercially available kits provide high-performance elution of DNA, contain enzyme (such as DNase) inhibitors and allow getting rid of impurities. An important advantage in NGS is the ability to simultaneously sequence many samples. This is done by marking samples by specific, short DNA fragments of known sequence treated as barcodes. The principle of the sequencing uses fluorescently labeled nucleotides. During the attachment of one nucleotide, generation of a light signal occurs and the reaction is temporarily blocked. After registering the signal, a fluorescent label is cleaved enzymatically allowing the entry of the next nucleotide. Each of the nucleotides (A, T, C, G) has a different type of fluorescent label with unique excitation and emission wave- length. DNA is immobilized on the surface of the flow cell, which allows direct and equal access of polymerase to each of the DNA molecule (Hodkinson and Grice, 2015). At a distance of less than one micron, there are more than a thousand copies of the same DNA fragments to form one cluster. Different DNA fragments form separate clusters, allowing for simultaneous sequencing of millions of DNA fragments. The parameters of current devices are extremely high. Within 24 H, around 5 Gb (giga bases) of reads can be determined, when reading

200–300 bp fragments (V3-V4 hypervariable regions for example). With exceptionally large genomic projects, the device with the highest performance (HiSeq series) can be operated allowing up to 1 Tb of data to be generated within a few days (Illumina, 2016). Next-generation sequencing in combination with other molecular methods (including DGGE) is a very complex and indispensable method of testing microbiomes and the ecological biome. Metagenomic approach to gain knowledge of the biodiversity present in difficult conditions, such as contaminated soil or sewage, sidelines all other known methods, allowing the examination of not only the whole fraction of microorganisms, but also discovering new, previously unknown species (Samarajeewa et al., 2004; Gorgé et al., 2016; Frisli et al., 2013).

Using the Flow Cytometry Method to Determine Microbial Diversity

Flow cytometry is an extremely versatile tool that complements existing technologies and enables fundamentally new information to be obtained in microbial ecology studies (Leglize et al., 2008; Gałązkaand Gałązka, 2015; Barac et al., 2004; Narasimhan et al., 2003; Aitken et al., 1998; Bumpus, 1989; Galazka et al., 2010; Theron and Cloete, 2000). Flow cytometry used in conjunction with fluorescence-activated cell sorting (FACS) can quantify and fractionate complex bacterial communities (Leglize et al., 2008). Typically, 3,000–100,000 cell/s can be characterized and sorted by flow cytometry. The type of information that can be obtained using flow cytometry includes size (Galazka et al., 2010), shape, surface texture, viability, DNA content (Cheem et al., 2010), and Specific staining technique based on fluorescent antibodies (Galazka et al., 2010) or rRNA-targeted oligonucleotide probes (Galazka et al., 2010; Yrjala et al., 2010) is used. Moreover, the sorting criteria used in FACS can be adjusted to allow fractionation of bacterial population based on any one, or a combination, of these parameters and the number of cells having certain characteristics as measured by flow cytometry can be quantified. Multiple fluorescent dyes are available that can selectively stain lipids, proteins, nucleic acids, and other cellular components (Narasimhan et al., 2003; Romero et al., 1998; Galazka et al., 2010; Theron and Cloete, 2000).

Using the Matrix-Assisted Laser Desorption/Ionization Time-of-Flight Mass Spectrometry (MALDI)-Based Method to Determine Microbial Diversity

Another automated system identifies microorganisms by determining the specimen mass spectrum and then comparing it to a database that contains known mass spectra for thousands of microorganisms. This method is based on Matrix-Assisted Laser Desorption/Ionization Time-Of-Flight mass spectrometry (MALDI-TOF) and uses disposable MALDI plates on which the microorganism is loaded with a specialized matrix reagent. The sample/reagent mixture is irradiated with a high intensity pulsed ultraviolet laser, resulting in the ejection of gaseous ions generated from the various chemical constituents of the microorganism. These gaseous ions are collected and accelerated through the mass spectrometer, with ions traveling at a velocity determined by their mass-to-charge ratio (m/z), thus, reaching the detector at different times. A plot of detector signal versus m/z yields a mass spectrum for the organism that is uniquely related to its biochemical composition. Comparison of the mass spectrum to a library of reference spectra obtained from identical analyses of known microorganisms permits identification of the unknown microbe (Spitaels et al., 2016).

Chapter 5

Extracellular Polymeric Substances and Corrosion

Abstract

Microorganisms on metallic surfaces can have a long-lasting effect on materials. Surface-associated microbial growth, a biofilm, is known to promote interfacial physicochemical reactions that are not favored under abiotic conditions. In case of metallic materials, undesirable changes in material properties due to formation of a biofilm are referred to as biocorrosion or microbially influenced corrosion (MIC). Microbially produced extracellular polymeric substances (EPS), which comprise of different macromolecules, mediate initial cell attachment to the material surface and constitute a biofilm matrix. Despite their importance in biofilm development, the extent to which EPS contribute to biocorrosion is not fully understood. Extracellular polymeric substances comprise the interaction between metal ions as well as anions such as pyruvate, succinate, glycerate, carboxyl, phosphate and sulphate for the metal binding. Characterization of metal and EPS interactions can be done by different techniques.

Keywords: extracellular polymeric substances, microbes, SEM, biofilm, environment

Introduction

EPS: An Overview

Extracellular polymeric substances (EPS) play a vital role in the formation of biofilm, cell adhesion as well as in cell aggregation and protect the cells from unfriendly environment. (Dogsa et al., 2005; Karn et al., 2017). Extracellular polymeric substances are the polymers which are biosynthesized by various microbial strains primarily due to the environmental signals and is composed of polysaccharides, proteins and DNA as per environmental conditions. The production of these extracellular polymeric substances, cause slimy layer

(Flemming and Wingender, 2010). In extracellular polymeric substances, polysaccharides form the most studied component and present in varied structure and composition (Roca et al., 2015). Further, the interest in extracellular DNA, enzymes and structural protein has also increased apart from the carbohydrates (Figure 5).

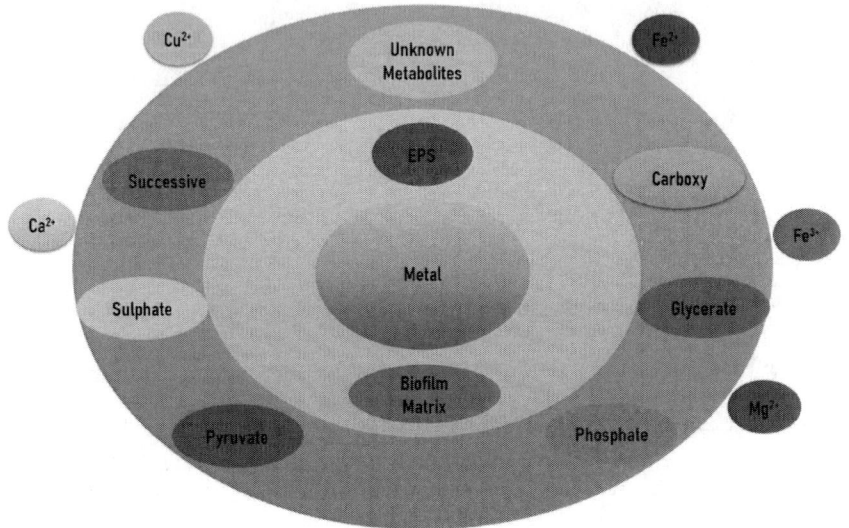

Figure 5. Mechanism of EPS in corrosion induction.

Extracellular polymeric substance forms matrix around the bacterial cells that shields them against heavy metals and antimicrobial compounds. The matrix also serves to retain their intrinsic environment against the drought, water and protecting microbes. Furthermore, other functions have also been reported like aggregation, adhesion, storage of carbon, entrapment of nutrients, communication with other microbes and plants, antioxidant (Vardharajula and Ali, 2015; Wang et al., 2015).

Extracellular polymeric substances are biosynthetic polymers that is mainly composed of lipids, polysaccharides, structural proteins, enzymes, nucleic acids and also some other compounds like humic acids (Flemming and Wingender, 2010). These EPS can also be produced by yeasts (Pavlova and Grigorova, 1999), protists (Jain et al., 2005; Lee Chang et al., 2014), fungi (Hwang et al., 2004; Elisashvili et al., 2009), microalgae, bacteria and cyanobacteria, (Parikh and Madamwar, 2006; Boonchai et al., 2014). The primary component of biofilm is the extracellular polymeric substances and approximately 50%–90% of organic carbon is present in the form of biofilm i.e., extracellular polymeric substances. Extracellular polymeric substances matrix has several physical as well as chemical properties. Due to the presence of various acidic materials like ketal-linked pyruvates and uranic acids, the contribution of polysaccharides to the ionic characteristics (or electric characteristics) of the biofilm, generally neutral to the anionic property and shows that many biofilms are electrically neutral to negatively charged. When the biofilms get develop these characteristics causes strong ionic bond with divalent cations like magnesium as well as calcium (Flemming, 2000; Sutherland, 1985) whereas some of the gram-positive bacteria demonstrate cationic charge property such as Staphylococci.

Commonly biofilms constitute non-uniform structure. The initial formation of biofilm is determined by the composition of polysaccharides. The solubility and the rigidity of the extracellular polymeric substances are determined by several polysaccharides 'structures such as 1,3- or 1,4-β-linked hexose causing less soluble and more rigid extracellular polymeric substances. The production of different amounts of extracellular substances may be helpful in characterizing different microorganisms. Moreover, the amount of extracellular polymeric substances produced is affected by the age of biofilm such that more amount of extracellular polymeric substances has-been produced as biofilm get older. Bacterial growth rate may get adversely impacted on the production of EPS due to the desiccation of biofilm (Sutherland, 1985). These extracellular polymeric substances have a unique nature of hydrophobicity. Mostly, they are highly hydrated as they have significant amount of water molecules. Nevertheless, in specific conditions, it can be both hydrophilic as well as hydrophobic in nature.

Source of EPS

Microbial extracellular polymeric substances find application in a wide range of industries such as textile, pharmaceuticals, food, cosmetics etc. as they are environmentally friendly polymeric materials. (Khani et al., 2015). While traditional EPS derived from other natural sources such as animals, plants and algae may fail to perform in some applications, bacterial EPS seems to have improved properties. Moreover, bacterial production is highly productive and less resource-intensive as compared to synthesis of extracellular polymeric substances by higher algae and plants.

Furthermore, to obtain the desired properties as well as higher yields, the production by bacteria enables control of process conditions (Alves et al., 2009). However, commercially microbial extracellular polymeric substances may be insufficient produced due to their high production costs like gellan gum, xanthan as well as hyaluronic acid. Hence, it is essential to make the process more cost-effective that can well be achieved through development of high yield strains, optimization of culture conditions genetic manipulation and low-cost substrate applications (Freitas et al., 2011). The synthesis and the composition of EPS is affected strongly by the nutritional and environmental conditions (Kumar et al., 2007). Generally, sources of nitrogen and carbon play a vital role as these nutrients are linked directly to the cell proliferation and metabolite biosynthesis (Kim et al., 2003; Lopez et al., 2003). Likewise, the nature and concentration of carbon sources are regulated by secondary metabolism via phenomena such as catabolic repression (Görke and Stülke, 2008).

Bacterial exopolysaccharides are composed of xanthan, glucuronan, cellulose, curdlan, succinoglycan, colanic acid, alginate, dextran etc. There are some microbes which produces exopolysaccharides, viz. *Xanthomonas* sp., *Pseudomonas aeruginosa, Escherichia coli, Salmonella* sp., *Lactobacillus helveticus, Acetobacter* sp., *Rhizobium meliloti, Shigella* sp., *Lactobacillus rhamnosus, Enterobacter* sp. (Ahmad et al., 2015). Several microbes such as *Geobacillus thermodenitrificans, Bacillus thermantarcticus,* are isolated from the extreme environments including Antarctic ecosystems, geothermal springs, deep-sea hydrothermal vents and saline lakes and studied possible sources of exopolysaccharides (Freitas et al., 2011). Extracellular polymeric substances have high molecular weights of about 10 to 1000 kDa and are either homopolymeric or heteropolymeric in composition (Nwodo et al., 2012). Usually, EPS contain carbohydrates and proteins besides nucleotide, humic substances and lipids (Fleming, 2000). The characteristics of bacterial extracellular polymeric substances are recognized on the basis of their molecular structure, chemical composition, average molecular weight, conformation of single molecules and their assemblies in case of aggregation process as well as gel systems (Morris et al., 2011; Iijima et al., 2007; Simsek et al., 2009).

Exopolysaccharides are produced by the bacteria that are generally present in a diversity of ecological niches. High ratio of nitrogen to carbon in the effluents from the paper, food and sugar industries and sewage treatment plants facilitates growth of such microbes (Morin, 1998). They also thrive in terrestrial and marine hot springs. Several microorganisms produce extracellular polymeric substances by the capsular material or isolated by the slime (Bajaj et al., 2007). These microorganisms have various genera of bacteria (mesophilic, thermophilic, halophilic), fungi and algae. Lactic Acid Bacteria is a famous mesophilic bacterium while there are other bacteria like *Aureobasidium* sp., *Acetobacter* sp., *Escherichia* sp., *Pseudomonas* sp., *Streptococcus* sp., *Escherichia* sp., *Lactobacillus* sp., *Bacillus* sp. etc. *Thermococcus* and *Archaeoglobus fulgidus Sulfolobus* are thermophilic archaebacteria (Nicolaus et al., 1993; Rinker and Kelly, 1996; Lapaglia and Hartzell, 1997). Many of the thermophilic bacteria produce excellent extracellular polymeric substances such as *Bacillus thermantarcticus, Geobacillus thermodenitrificans and Bacillus licheniformis*. Huge amounts of EPS are produced by the co-cultures like *Methanococcus jannaschii, Thermotogamaritima* and *Geobacillustepidamans V264* (Kambourova et al., 2009). Various halophilic archaea such as *Natronococcus*, *Halobacterium, Haloarcula, Haloferax, Halococcus* also generate EPS (Antón et al., 1988; Nicolaus et al., 1999; Paramonov, 1998).

It has been asserted that that metal biocorrosion process involves biofilm-related anaerobes just like planktonic and sulfate-reducing bacteria (Beech and Sunner, 2004; Hamilton, 1985). It has been known that the sulfidogenic and acidic metabolic products can damage the oil pipelines and infrastructure, unintended products are released and deteriorate pipelines through the metal loss (Kilbane et al., 2005). Many microorganisms like Sulphur-reducing bacteria exists as biofilms or cell aggregates and surrounded by extracellular polymeric substances matrix. Branda et al., (2005); Flemming and Wingender, (2010); Wingender et al., (1999) describes that amongst their various functions, EPS provide cell protection, bacterial nutrition source, allow for the stabilization of biofilm as well as substratum attachments. Biofilm-associated cells produce extracellular material for the structural integrity of cells (Flemming and Wingender, 2010; Kolter and Greenberg, 2006).

Overall, the extracellular material that accounts for more than 90% of the dry weight of a biofilm, comprises a combination of bacterial metabolites, proteins, nucleic acids, modified polysaccharides and lipids all of which regulate the matrix architecture as well as its interaction with a surface or substratum (Branda et al., 2005; Flemming and Wingender, 2010; Wingender et al., 1999; Sutherland, 1985).

Numerous marine hydrocarbon-degrading microbes have amphiphilic extracellular polymeric substances which play an essential role in solubilizing and biodegradation and bioavailability of hydrocarbon compounds, a characteristic that may be of possible biotechnological use in response to oil spills such as *Halomonas* genus members that breakdown selected polyaromatic as well as aliphatic hydrocarbons (Gutierrez et al., 2013). The EPS which solubilize the polyaromatic hydrocarbons also combine several hydrocarbon-containing compounds such as crude oil, n-alkanes, food-grade oils; moreover, for biodegradation, it enhances the bioavailability of hydrophobic materials, too (Gutierrez et al., 2013; Gutierrez et al., 2007). Frequently, these capabilities are accredited to higher amounts of uronic acid as well as peptides in *Halomonas*-produced EPS which accounts for their amphiphilic nature (Gutierrez et al., 2009). Remarkably, the presence of EPS from the *Halomonas* strain TG39, enhances the biodegradation of phenanthrene by a deep-water horizon bacterial community, signifying that EPS helps to facilitate the bacterial access to probable carbon substrates in oil droplets (Gutierrez et al., 2013). EPS of non-hydrocarbon-degrading Sulphur reducing bacteria such as *Desulfovibrio* sp. are also notable among EPS producing microbes. Two *Desulfovibrio* strains (Ind1 and Al1, later classified as *D. indonesiensis* (Feio et al., 2004) and *D. alaskensis* G20 (Hauser et al., 2011) respectively produces exopolymers isolated from the marine corrosion failures comprise amino sugars, neutral hexoses, nucleic acids, protein as well as uronic acids (Zinkevich et al., 1996). Furthermore, extracellular polymeric substances of *D. indonesiensis* Ind1 strain comprise an iron-chelating carbohydrate-protein complex that is corrosive to mild steel (Beech et al., 1998).

EPS in the Corrosion Process

The binding capability in extracellular polymeric substances to the metal surface in microbially influenced corrosion depends both on type of metal as well as the species of microorganism (Beech and Coutinho, 2003; Kinzler et al., 2003; Sand, 2003; Rohwerder et al., 2003). Extracellular polymeric substances comprise the interaction between metal ions as well as anions such as pyruvate, succinate, glycerate, carboxyl, phosphate and sulphate for the metal binding. These ions get bound with super-molecular backbones consisting of polysaccharides (Sutherland, 2001). Precisely, for multivalent ions like Mg^{2+}, Cu^{2+}, Ca^{2+} and Fe^{3+} the affinity of multidentate anionic ligands remains quite strong.

The presence of metal ions innumerous oxidization states within the biofilm matrix and the standard reduction potentials may end up in substantial shifts, even though the presence of metal ions at intervals in the biofilm matrix has been recognized as relevant to MIC. The possible involvement of EPS-bound metal ions in direct electron transfer from the base metal to an acceptable electron acceptor involves ferrous metal. Matrices furthermore have a significant influence on the activity levels of biofilm organisms present within them. The oxidation of metal ions adsorbable on the surfaces of oxyhydroxide iron that assembles on biofilm polymers probably further contributes to cathodic reactions.

It is exciting to think about such biominerals having electron-conducting fibres distributed at intervals in the biofilm matrix (Beech, 2004). Beech and Sunner, (2004) reported metal ions, bound inside the biofilm matrix, to be coordinated by a spread of ligands, thereby forming complexes presenting a spread of redox potentials. Such complexes might participate within the electron transfer processes that drive corrosion reactions. The importance of such a mechanism within the overall corrosion method powerfully depends on the chemistry of the metallic surface associated with each bacterium and is specific to each bacterium species (Busalmen et al., 2002). Charge on polysaccharides may be associated with either carboxyl groups of uronic acids or with non-carbohydrate substituents (Sutherland, 2001). Proteins rich in acidic amino acids, as well as in aspartic and aminoalkanoic acid, bear carboxyl groups that additionally contribute to the anionic properties of EPS. Nucleic acids are polyanionic to the phosphate residues within the ester moiety. Thus, charged elements of acidic amino acids, EPS, uronic acids and phosphate containing nucleotides are likely to be involved in electrostatic interactions with multivalent cations.

Dong et al., (2011) studied the effects of extracted EPS on steel corrosion. They discovered that some types of EPS delayed the corrosion by inhibiting oxygen ion mass transfer. However, corrosion inhibition decreased at a high EPS concentration because of the binding of additional iron ions to EPS. Once microorganisms colonize a metal surface, they tend to produce EPS that contains very different functional groups. These changed functional groups now have totally different affinities for the metal ions, which leads to zones of changed metal concentration. The area found below the surface exopolymers showing high affinities for metal act as anodes, whereas the areas underneath exopolymers with low metal affinities act as cathodes. Bautista et al., (2015) determined that EPS produced by bacteria plays a primary role in the different stages of microbial biofilm formation, maturation and maintenance. The different influences were found in loosely bound (LB) and tightly bound (TB) EPS, respectively, isolated from the marine bacterium *Pseudomonas* (NCIMB, 2021). This role may be compared with that played in the presence of the model protein bovine serum albumin (BSA). A thick duplex chemical compound layer deposited over a Cu_2O layer and an inner oxidized nickel layer of thickness of order 10 nm (Conlisk, 2013; Jenkinson, 2014) showed the presence of BSA, TB EPS and LB EPS, forming a mixed chemical compound layer (oxidized copper and nickel) of layer thicker than that of the underlying layer of 70Cu-30Ni alloy in static artificial seawater (ASW) but not thinner than a layer of biomolecules. In solutions of biomolecules, this oxide layer is covered by an adsorbed organic layer that is composed principally of proteins. Wikieł, (2013) used a model to examine impedance values obtained at the corrosion potential and showed a slowdown of the anodic reaction in the presence of biomolecules (BSA, TB EPS and LB EPS), as well as corrosion inhibition by LB -EPS and to a lesser extent by BSA.

Techniques to Determine EPS

EPS interactions vary depending on its composition, metal species, and environmental conditions. Characterization of metal and EPS interactions can be done by different techniques like Polarographic titration method, FTIR spectroscopy, three-dimensional excitation-emission matrix (EEM) fluorescence spectroscopy, scanning electron microscopy (SEM), energy dispersive X-ray (EDX) microanalysis, nuclear magnetic resonance spectroscopy (NMR) and thermodynamic analysis (Li and Yu, 2014).

Chapter 6

Proteins in Corrosion

Abstract

A variety of proteolytic as well as hydrolytic enzymes has been produced by the bacteria which reacts with the substrates as well as cell wall and are known as exoenzymes. There are various enzymes such as glycosidases, polysaccharides, oxidoreductases, esterase, proteases, phosphatases, lipases and peptidases. Certain types of interactions between the substrates and cells are enabled by the discharge of enzymes into their ambient environment. In microbial influenced corrosion studies, the use of purified enzyme is identified as an indispensable tool to discover the chemical behavior of surface science processes mainly the ennoblement of stainless steel in waters. Mainly, glucose oxidase, peroxidase, catalase play major role in corrosion process. The increasing interest of researchers has been attracted by such ennoblement as the pitting corrosion which affects the economy and their essential consequences are still not clearly understood.

Keywords: proteins, amino acids, enzyme, hydrogen peroxide, manganese oxide

Introduction

Proteins: An Overview

Proteins are macromolecules and the polymers of amino acids. There are oddly 20 different amino acids that take part in the protein formation; hundreds of amino acids get covalently attached to each other in a long chain to form the working macromolecule. Due to the hydrolysis, amino acids get released from proteins. Obligatorily, proteins form colloids due to their large size when get dispersed into a suitable solvent and this property of protein characteristically distinguishes it from the solutions that contain small molecules (Blanco and Blanco, 2017). In animals, proteins are the most abundant organic compounds and composed of C, N, H, S and O.

Proteins can be classified according to the characteristics of their carbon chain. Amino acids contain both acidic as well as basic groups because of their amphiphilic nature. All the amino acids have asymmetric carbon except glycine exhibiting optical isomerism. All the amino acids belong to L-form in the animal world. The peptide bond links the carboxyl group of amino acid with the α-amine of α-C of another. The polymers of amino acids having less than 6000KDa are referred to as peptides whereas proteins are polymers of more than 6000KDa. According to Blanco and Blanco, (2017) in peptide, amino acid and protein, the dissociation of both positive as well as negative groups is same at isoelectric point and the overall net charge of molecule is zero.

Enzymes: An Overview

Cell is the basic building block of living systems and the structural and functional unit of life. Enzymes are effectively utilized by the cells and have an amazing catalytic efficiency as well as substrate and reaction specificity. They are vital for cellular metabolism; the chemical reaction occurs in animals, plants and micro-organisms proceeds at a quantifiable rate as a direct result of enzymatic catalysis.

Except the RNAse, all enzymes are active proteins which catalyse the biochemical reactions. These biomolecules are required for the breakdown and synthesis of reactions by the living organisms. These enzymes help to maintain and build all living organisms; capable of converting a substrate into the product at higher reaction rates. Enzymes enhance the rate of reaction like chemical catalysts by lowering the activation energy (E_a), so as to rapidly reach the equilibrium state. It has been found that the rate of enzymatic reactions is millions of times faster than that of uncatalyzed reactions. Transformation that would have taken hundreds of years can be executed in seconds (Patel et al., 2017). According to Patel et al., (2017) enzymes are known as enantio selective catalysts that uses either in the chiral compound synthesis or in the separation of enantiomers from a racemic mixture.

Role of Proteins and Enzymes in the Corrosion Process

A large variety of proteolytic as well as hydrolytic enzymes known as exoenzymes such as lyases are produced by bacteria which react with the substrates beyond the cell membrane as well as cell wall. Free forms of enzymes are those that are produced by the cells but acts outside it such as glycosidases, polysaccharides, oxidoreductases, esterase, proteases, phosphatases, lipases and peptidases. These enzymes are released into their immediate environment and act on the available substrates (Figure 6). In some marine biofilms, enzymes are of concern mostly as their influence needs to be detected (Basséguy et al., 2004; Rusling, 1998). According to Lai et al., (1999), these enzymes possess a haem group that are responsible for the electron transfer during the accelerator chemical process. There are several protein-driven mechanisms involved in the aerobic corrosion such as interaction with the porphyrin compounds as well as catalase whereas with hydrogenases, SRB and FeS in anaerobic corrosion. Amaya and Miyuki, (1995, 1999); Dupont et al., (1998); L'Hostis et al., (2003) believes that oxidase-type enzymes involve in the ennoblement of stainless steel. Generally, to catalyze the oxidation of organic compounds these enzymes use oxygen as an acceptor. Within the biofilm, reactive oxygen species has been produced as a consequence of such accelerator activities. Hydrogen peroxide (H_2O_2) is the reactive oxygen species that plays a vital role in various accelerator reactions occurring in aerobic biocorrosion (Amaya and Miyuki, 1999; Lai et al., 1999).

Essential involvement of H_2O_2 in stainless steel ennoblement has been correlated by number of studies in natural water because of its redox potential over the molecule of oxygen (Chandrasekaran and Dexeter, 1994; Washizu et al., 2004). Additionally, in corrosion, the enzymatic activity within the biofilm is thought to be involved (Beech, 2003; L'Hostis et al., 2003) whereas there are some researchers who have explained about the mechanism of stainless steel in oxygenated natural water which enhance the free potential of corrosion where it is well-thought that the ennoblement enzyme that are involved can be either intracellular or extracellular or both. Landoulsi et al., (2008) describes how the consumption of oxygen influences its availability through the metabolic pathways within the biofilm at metal-biofilm interface. Consequently, the gradients of oxygen result from its diffusion inside the biofilm and from their consumption in the metabolic pathways. In the microorganisms, oxygen acts as final electron acceptor and this oxygen undergoes four-electron reduction pathway. Cytochrome c enzyme is intracellular nevertheless some of the secreted forms are inducible in response to the oxidative stress (Naclerio et al., 1995). Catalase enzyme activity generates O_2 that can be used to produce H_2O_2 by the reduction on the surface of stainless steel. Busalmen et al., (1998, 2002) considered this as autocatalytic mechanism for the first time. In cathodic current, they discovered rise by using the *Pseudomonas* sp. culture which secrete the enzyme catalase. Primarily on a modified stainless-steel electrode, the enzyme immobilization-based technique has been developed by Landoulsi et al., (2008). In a polymeric film, the immobilized glucose oxidase spread on the surface of stainless steel to concentrate the activity of enzyme close to the metal as well as polymer film interface.

To check the impact of GOx, an electrochemical test has been performed with free or immobilized forms inside the electrolyte. These experiments show oxygen as well as hydrogen peroxide reduction and each of the oxidant involves inside the GOx-catalysed reaction. Large variations have been shown by the cathodic processes between the immobilized or freely associated proteins while the OCP ennoblement has been found because of electrochemical effects of hydrogen peroxide in each of the cases.

The strong oxygen depletion is resulted by immobilized GOx near the interface of polymer or metals. Because of the decrease in hydrogen peroxide, the reduction of current was recorded. Microbial enzymes catalyze the oxidation-reduction reactions (Scotto et al., 1985) of cellular chemicals (Mollica et al., 1990, 1992) and organometallic complexes (Johnson and Bardal, 1985). Chandra Sekaran and Dexter, (1993) described that at low hydrogen ion concentrations the peroxide that generated from the activity of microorganism contributes to ennoblement. The biofilms increased the passivity of SS alloys by each increasing the passive region and therefore the vital breakdown potential. Maruthamuthu et al., (1996) and Dexter and Zhang, (1991) described that due to the layer of n-type semiconducting chemical compound film, only alloys cover and exhibited a significant positive shift of corrosion potential on the basis of photoelectrochemical studies nevertheless, whether the biofilm improves passivity is still unknown.

Dickinson and Lewandowski, (1996) compare the quality of fresh natural water to seawaters and recognized that correlation exists between the bacterial deposition as well as ECORR ennoblement of manganese hydroxides or oxides. For the production of pitting corrosion of stainless-steel sample, manganese biomineralization is liable and exposed in low chloride media (Geiser et al., 2002; Shi et al., 2003).

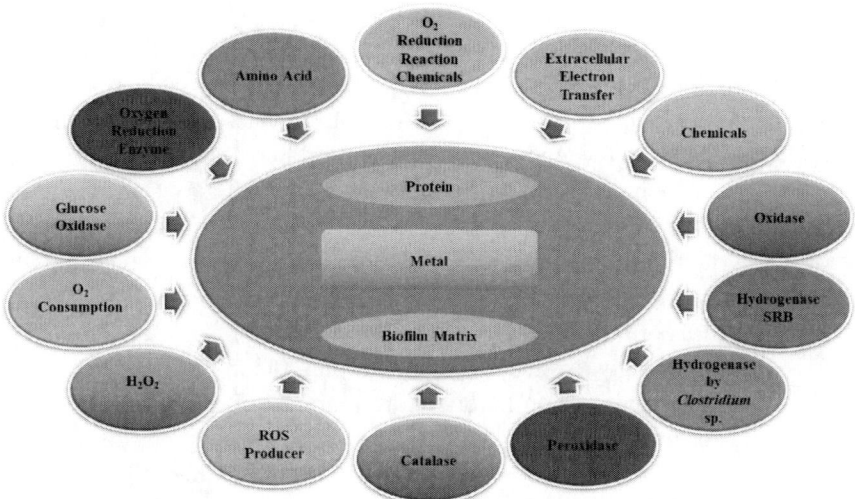

Figure 6. Protein mechanism in corrosion induction.

In the river Seine (France), natural exposure of stainless-steel sample both manganese as well as hydrogen peroxide biomineralization mechanisms has occurred depending upon the exposure site (Marconnet et al., 2008).

The ennoblement of AISI 316L has been reported by Landoulsi et al., (2009) that has been elicited when the GOx catalyzes a reaction or add up hydrogen peroxide in the seawater. The corrosion behavior of the sample has been studied by using the galvanostatic as well as potentiodynamic polarization tests. The pitting potential has become nobler once the ennoblement has occurred. For these processes, the involvement of hydrogen peroxide in reinforcing pit repassivation is a key component. In microbial influenced corrosion studies, the use of refined enzyme is progressively identified as an indispensable tool to uncover the chemical aspects of surface science processes mainly the ennoblement of stainless steel in the natural waters. The interest of many researchers has been attracted by such ennoblement as the pitting corrosion economically essential consequences which are still not clearly understood. The catalytic effect of biofilms has been reported after several studies with the enzyme involvement as well as their influence on the native physiochemical conditions related to their heterogeneous structure. Amaya and Miyuki, (1995); Chandershekhran and Dexter, (1993) and Washizu et al., (2004) described in stainless steel ennoblement, the involvement of hydrogen peroxide in natural water via its oxidoreduction potential beyond the oxygen. The presence of various binding sites has been included in the chemical properties of biofilm among the macromolecules for the formation of matrix. Enzymes promote close association between exogenous substrates as well as exo-polymeric substances and enable the biochemical rection. In waste material of biofilms, the 3D mapping of oxygen element distribution has been used which shows the pockets in deep parts of biofilm along with dissolved oxygen (De la Rosa and Yu, 2005) which enables the biofilm development. Deep inside the biofilm, certain dissolved oxygen has been used for their formation as well as maintenance partly by facilitating at least few of the metabolic activities deep in biofilms of huge mass.

Most reports on watching accelerator expression in biofilms specialize in cell-associated activities and barely address noncellular areas of the biofilm matrix. According to Beech, (2002) hydrogenases enzymes of anaerobic sulphate reducing bacteria remain active within the biofilm matrix irrespective of the absence of living cells and play a vital role in iron as well as metal alloys biocorrosion. Furthermore, in aqueous oxygenated solutions of freeze-dried sulphate-reducing bacteria exopolymer can be detected by the activity of enzymes such as esterases, phosphatases, lipases and catalases (Beech, 2002; Beech and Coutinho, 2003) but the enzymatic impact on steel corrosion is not clearly understood.

These studies indicate the wide physiological versatility of various microorganism species and have pointed to considerable species specificity, which may lead to the apparently paradoxical diversity of deterioration processes observed in biofilms even under apparently identical environmental conditions (Dinh et al., 2004). At single-cell level, the close observation of microorganisms that processes on surface is needed to resolve in the apparent paradox of biodiversity as has been done for plankton and resolved correspondingly by Hutchinson, (1961) and Tilman, (1982).

The enzymatic activity plays a vital role in the dissolution of hydroxide or oxide films coating as well as protecting the surfaces of metal. Consequently, on the surface of steel, the passive layers will be replaced or lost by the less reduced or stable metal film which facilitate the corrosion. Gram-negative bacteria such as *Shewanella oneidensis* (formerly known as *S. putrefaciens*) are the most effective example of microorganism which produces the biocorrosion due to the dissimilarity in the reduction of iron. Various substrates of carbon have been oxidized by this bacterium by reductively dissolving Fe (III) present in minerals such as haematite, ferrihydrite as well as goethite. Lee and Newman, (2003) suggested the presence of *S. oneidensis* in biocorrosion. The rate of corrosion is measured on the basis of type of oxide film under the attack (Little et al., 1997).

Anaerobic biocorrosion mechanisms have been projected by the SRB and TRB. The catalysis of proton reduction to molecular hydrogen occurs after the precipitation of iron compounds. According to Munoz et al., (2007), hydrogen acts as a cathode for galvanic couple with iron, bronze as well as other anodic materials with depolarization promoting local acidification at the anode possibly the corrosive phosphides production releasing matter such as PH3 that enhance the water dissolution under the anaerobic conditions (Iverson and Olson, 1983). Beech and Cheung, (1995) and Miranda et al., (2006) describe that by further cellular polymer substances, metal ion complexation reduces the pH by allowing the metabolic reduction of thiosulphate to sulphide.

In the nitrogen fixing, methane-oxidizing bacterium, the hydrogenase activity arises once their growth depends upon the fixed gaseous nitrogen (De Bout, 1976). During the bacterial growth, a direct relationship exists between the hydrogenase as well as nitrogenase activity. Due to the presence of hydrogen gas, acetylene reduction is additionally supported. Organic acid has been produced by the fungal degradation of lubricating grease (Little et al., 2001). In polyvinyl chloride sheaths, localized corrosion of carbon steel has been produced. In the exceedingly post tensioned structure, *Penicillium* sp., *Fusarium* sp. and *Hormoconis* sp. has been isolated. In these cases, localized corrosion was observed as the spores of fungus get deliberately introduced to cover the steel tendons and there is a spatial relationship between the corrosion as well as fungal hyphae. It has been described that an antibacterial autolytic protein i.e., AlpP is produced by *Pseudoalteromonas tunicate,* a marine bacterium and causes death of a portion of cell population during the formation of biofilm and mediates the dispersal, phenotypic variation as well as differentiation among the dispersed cells (Prochnow et al., 2008). Currently, in the marine bacterium i.e., *Marinomona smediterranea*, a homolog of AlpP (LodA) has been identified as a source of lysine oxidase that mediates cell death via hydrogen peroxidase production, and this has been confirmed by Prochnow et al., (2008) that AlpP in *P. tunicate* acts as a source of hydrogen peroxidase and that lysine oxidase is responsible for causing cell death in microcolonies inside the *P. tunicate* and *M. mediterranea* growing biofilms.

The cell death distribution is varying the LodA-mediated that is related to phenotypically determined growth pattern and the formation of biofilm amongst the *M. mediterranea* biofilm dispersal cells. Moreover, in various different gram-negative bacteria, AlpP homologue occurs from the environments. In microcolonies, the subpopulation of cell death confirms the biofilm formation in *Chromobacterium violaceum* and *Caulobacter crescentus*. At the time of biofilm death, hydrogen peroxide involves in these organisms and detects when killing occurs.

Moreover, the catalase enzyme decreases the killing of biofilm. Since the isogenic mutant i.e., CVMUR1 that do not suffer in the death of biofilm in *C. violaceum*, the AlpP homologue connects with the death events of biofilm. The biofilm that connects with the AlpP homologue activity might be killed via peroxide and constitutes across a gram-negative microorganism spread and also plays a vital role in the diffusion. During the anaerobic corrosion of mild steel, the hydrogenase activity by *Clostridium acetobutylicum* has been found by Mehanna et al., (2008).

On single coupons, the effect of hydrogenase was confirmed by the free potential observation whether or not exposed to enzyme in an ideal cell when it is completely deoxygenated. The hydrogenase enhances the free potential around 60 mV in all Tris HCl as well as phosphate buffer and the iatrogenic effect marks the general corrosion. It has been stated by the researchers that in the absence of any final electron acceptor, the iron hydrogenases act by catalyzing the direct electron acceptor on the surface of mild steel. Davidova et al., (2012) demonstrated that from the inner surface material of hot oil pipeline, two thermophilic archaea i.e., strain PK and strain MG has been isolated from enriched culture at 80°C.

Complex organic nitrogen sources such as peptone, yeast extracts as well as tryptone has been fermented by the strain PK and may decrease the elemental sulphur like Mn^{4+}, S and Fe^{3+}. The phylogenetic analysis shows that organisms belong to the Thermococcales order. During the fermentation of yeast extract, ferrous iron is formed abiotically resulting in the incubation of this strain with elemental iron and consequently, volatile fatty acids take sometimes.

For the reduction of metal, an essential reaction that takes place in various anoxic environments on the microorganism is extracellular electron transfer to solid surfaces (Ishii et al., 2014). Nevertheless, in microbial communities, it is very difficult to characterize the extracellular electron transfer as well as contribution to each member of the community due to the changes in the composition and concentrations of electron donors as well as solid-phase acceptors. To measure the synergistic effects of surface redox potential as well as carbon sources on the extracellular electron transfer-active microorganism community development, complete electron transfer as well as metabolic networks by using bio-electrochemical systems. Under the electropositive electrode surface potential as well as fatty acid-fed conditions founds to be occurred in rapid biocatalytic rates. After the analysis of temporal microorganism community, it has been found that Geobacter phylotypes are tremendous and depends upon the surface potential. These microbes are associated with rather low current generation as well as low surface potential and affiliated with *G. metallireducens*.

For each solid surface, these collective results demonstrated that at species as well as strain level, the surface potential gives a powerful selective pressure for fermentative and respiratory bacteria throughout the extracellular electron transfer-active community. Da Silva's et al., (2004) discovery of fluctuation between positive and negative values suggests that the hydrogenase enzyme favours both anodic and cathodic sites on an equivalent conductor. Its global behaviour is controlled by the balance between the native anode and cathode sites. Such native anode/cathode sites on steel surfaces already have been induced in the presence of hydrogenase from *Ralstoiniaeutropha*. Advanced carbon substrates are degraded by fermenting microorganisms to by-products of the fermentation methods. Such by-products include volatile fatty acids and H in an anaerobic system which is utilized as electron donors for sequential microbial reduction of nitrate and sulphate, as well as reduction of solid metals through microbial EET reactions (Lovley et al., 2004; Nealson and Saffarini, 1994). The high rates of microbial electron uptake discovered through microbially influenced corrosion of iron Fe (0) and through microbial electrosynthesis have been thought to indicate immediate electron uptake in these microbial processes. However, the molecular mechanisms of direct electron uptake from Fe (0) remain unknown.

Deutzmann et al., (2015) investigated the facilitation of electron uptake characteristics of Fe (0) corrosion by extracellular enzymes through *electromethanogenic Archaea* and *Methanococcus maripaludis*. The authors' results suggest that the free, surface-associated redox enzymes including hydrogenases and formate dehydrogenases mediate direct electron uptake. Rates of H_2 and formate formation by a cellular spent medium were used to elucidate rates of methane formation from Fe (0) and cathode derived electrons by wild-type M. maripaludis, as well as by a mutant strain carrying deletions for all catabolic hydrogenases. Amino acids areas are characterized by eco-friendly properties such as water solubility, natural occurrence and biodegradability (Morad, 2005). Amino acids have been reported as corrosion inhibitors for Cu, Al, Sn and Fe, with inhibition related to the nature of both the metal and the medium (Ece and Bilgic, 2010), particularly amino acids containing sulphur and long organic chains (Mobin et al., 2016). Much research has targeted corrosion inhibition by amino acids on copper (Mobin et al., 2011) and iron (Fu et al., 2010). The degree of inhibition by the medium can be related to its composition of amino acids and in some cases is increased by addition of certain cations and anions (Migahed and Al-Sabagh, 2009).

Recent enzymatic activities, measured per unit area in biofilms, have shown enzyme-mediated reactions to be more important than was previously thought and are thus likely to be highly relevant to biocorrosion. The mechanism, which increases the free corrosion potential (ECORR) of SS, has been widely reported in oxygenated natural waters and cited as an ennoblement agent, as our understanding improves (Beech, 2003; Busalmen et al., 2002). In microorganisms, enzymes, including catalases, peroxidases and superoxide dismutase, are part of the respiratory chain and act as accelerators. They are involved in oxygen reduction. Thus, they may facilitate corrosion by accelerating oxygen reduction reactions. However, it is imperative to comprehend that the power of such enzymes to accelerate oxygen reduction depends strongly on the chemistry of surface films (Beech, 2004).

Recently, Awad et al., (2017) examined the inhibition of corrosion of mild steel in the presence of various amino acids, including cysteine, methionine and methyl-cysteine complex. The analyses were done with each amino acid alone and also in combination. The most marked inhibition was found in the presence of amino acid mixtures. Many factors and processes in biofilm development can lead to reduction in corrosion (Jayaraman et al., 1999). Earlier Chongdar et al., (2005) discovered aerobic *P. cichorii* had the potential to inhibit corrosion of mild steel because of a layer of passive oxide products formed by corrosion, as did Dubiel et al., (2002) who found that microorganism respiration led to oxygen removal by biofilms, which they surmised explained the inhibition of corrosion. Similarly, Juzeliunas et al., (2006) found that biofilm produced by *Bacillus mycoides* magnified charge transfer resistance of the Al substrate and thereby decreased the corrosion rate.

Techniques to Determine Enzymes in Corrosion

The precise structure of biofilm allows the flow of waste products, nutrients, solutes, metabolites as well as enzymes (Costerton et al., 1994; Lappin-Scott and Costerton, 1995; Sutherland, 2001; Costerton and Boivin, 1991). The matrix of biofilm considered as immobilized enzymes system in which the activity of medium and enzyme are changing constantly as well as evolve to reach a steady state (Sutherland, 2001). Currently, in wastewater biofilms, the three-dimensional mapping of oxygen distribution has been observed (Rosa and Yu, 2005). Within the biofilm, some highly concentrated pockets of oxygen were revealed (Rosa and Yu, 2005). In the biofilm, by the use of glutaraldehyde, a bactericide treatment has been proposed to remove the consumption by microorganisms that form in the seawater (Lewandowski et al., 1989).

Oxygen acts as a final electron acceptor in the microorganisms and their major part undergoes four-electron pathway reduction in the microorganisms and catalyzed by cytochrome-c oxidase (EC 1.9.3.1). For enzymes, the EC number (Enzyme Commission number) is a numerical classification on the basis of chemical reactions. For the respective enzyme, every EC number associated with the recommended name for respective enzyme as a system of enzyme nomenclature (IUBMB, 1992). Oxygen is subjected to one-electron biological reductions. This may lead to the formation of radical or molecular intermediate products, called reactive oxygen species (ROS) because of their higher reactivity than the oxygen itself.

On metallic substrates, the decrease in oxygen is thermodynamically possible under the aerobic conditions. Salvago and Magagnin, (2001) described that even if the reaction is very slow, it can be catalyzed by various chemical species like organic as well as inorganic mediators (Polypyrrole and Fe^{2+} correspondingly). Decrease in the abiotic oxygen results from different pathways (Yeager, 1986): according to a single step, in four-electron mechanism or two successive electron reduction steps that is, with the intermediate formation of hydrogen peroxide (Okuyama and Haruyama, 1990; Yang and McCreery, 2000) as follows:

$$O_2 + 2e - f2H^+ + H_2O_2(2) + H_2O_2 + 2e^- f2H^+ + 2H_2O$$

The oxygen reduction mechanism on SS surfaces is strongly influenced by biofilms (Shams et al., 1996). In fact, prior to the formation of a biofilm, oxygen reduction occurs. After a while, the biofilm grows on the metal surface and ECORR ennobles to positive values.

Electrochemical tests combined with XPS (X-ray photoelectron spectroscopy) analysis allowed to show the important role played by SS passive films on oxygen reduction in natural seawater (Bozec, 2001). MIC studies evidencing the oxygen reduction catalyzed by biofilms have been performed using the most common electrochemical techniques (Holthe et al., 1989): (i) by registering dynamic cathodic polarization curves after ECORR ennoblement, starting either from ECORR value or from an anodic potential; (ii) by registering cathodic current as a function of immersion time in potentiostatic tests at the potential range where oxygen may be reduced.

Enzyme-Influenced Oxygen Reduction

The influence of enzymes has been recently investigated (Basséguy et al., 2004; Rusling, 1998): peroxidase (EC 1.11.1.7) and catalase (EC 1.11.1.6) are of particular concern, as they have been detected in some marine biofilms (Lai et al., 1999). All these proteins possess a heme-group (also called hemins) which is involved in the transfer of electrons during the enzymatic catalysis. When hemin alone was used in local electrochemical tests using SVET, the cathodic behaviour of the hemin-coated area and the anodic behaviour of the bare area of the SS surface were confirmed (Basséguy et al., 2004). Hem in alone has no enzymatic activity, but it proved its ability to shift ECORR toward anodic values when immobilized on an SS surface (Basséguy et al., 2004). Therefore, in proteins with no enzymatic activities, the presence of hemin is sufficient to enhance oxygen reduction on SS surfaces.

Pivotal Role of Hydrogen Peroxide

Biofilms formed in natural seawaters are able to generate hydrogen peroxide (H_2O_2) (Xu et al., 1998; Dickinson et al., 1996; Denis et al., 1989; Washizu et al., 2001), one of the reactive intermediates of oxygen reduction. The H_2O_2 concentration within the biofilm, estimated to be in the range of mmol/L, is governed by two antagonist processes: the production by some oxidases (enzymes using O_2 as electron acceptor) and the simultaneous degradation by enzymes which protect microorganisms against oxidative stress (catalases, peroxidases). In fact, many bacteria released hydrogen peroxide as the end product of oxygen reduction and contribute to the steady level of H_2O_2 in seawater (Denis et al., 1989; Murphy and Condon, 1984; Kairuz et al., 1988; Zika et al., 1985). From the electrochemical point of view, H_2O_2 is a common oxidant which exhibits a very high standard potential (E°) 1.77 V/SHE. Because of its redox potential higher than that of oxygen, H_2O_2 plays an important role in the ennoblement of SS in natural waters (Chandrasekaran and Dexter, 1993; Amaya and Miyuki, 1999; Dupont et al., 1998; Chandrasekaran and Dexter, 1990; Franklin and White, 1991; Washizu et al., 2004; Bozec et al., 1998).

Moreover, ferrous and ferric compounds are able to catalyze H_2O_2 decomposition (Zepp et al., 1992; Kremer, 1985). Direct role of H_2O_2 in the ECORR ennoblement is controversial, this oxidant is at the crossroad of many enzymatic reactions involved in aerobic MIC mechanisms.

Enzymes Producing H_2O_2: Oxidases

Dupont et al., (1997) suggested that EPS (exo-polysaccharides) alone were not able to reproduce the electrochemical effect of biofilms. This conclusion was obtained experimentally by monitoring ECORR at two different temperatures: 25 and 40°C.

Microbial-Induced Deposition of Mn Oxides

MnO_2 is formed from Mn^{2+} by Mn oxidizing bacteria, and then it is reduced. In all these approaches, we can highlight two important points: (a) the reduction of Mn oxides is a cathodic reaction which can be summed to the oxygen one, and therefore it explains ECORR ennoblement; (b) the electrochemical effect observed in the presence of Mn oxidizing bacteria requires a direct contact between Mn oxides/hydroxides and the SS surface. Because of the low growth rate of manganese oxidizing bacteria, they are easily removed from the SS surface by other microorganisms (Campbell et al., 2004). When this competition occurs, the contact between manganese-oxidizing bacteria and SS surface is lost and the deposition of MnO_2 is no longer possible.

Chapter 7

Role of Lipids in Corrosion

Abstract

Corrosion is a worldwide issue that extensively affects many industries. Most corrosion is the effect of electrochemical processes driven by microorganisms present within biofilms. It is essential to know the fundamentals of biocorrosion and also the roles of biomolecules within the whole process to develop detailed research capabilities and potential control and management.

This chapter targets the roles of lipids and different substances currently known to be involved in corrosion processes. The potential roles of lipids are still poorly understood and form the central subject of study in the chapter. More specific questions concern the impact of lipid activities inside the biofilm matrix on the dynamics of corrosion reactions.

Keywords: lipids, corrosion, solvent, ATP, electron transfer, liquid chromatography, gas chromatography

Introduction

Lipids: An Overview

Lipids are biomolecules having their high solubility in non-polar solvents. Lipid consists of only triacylglycerols which may contain a huge number of different chemical functional groups. It is often important for scientists to either know or to be able to specify the concentration of different types of lipid molecules present, as well as the total lipid concentration. Some of the most important reasons for determining the type of lipids present have been discussed below (Karn et al., 2020).

Role of Lipids in Corrosion Process

Basically, electron tunneling determines the chance of whether or not an electron can move from a donor to an acceptor molecule (Figure 7). Tunneling between donor and acceptor molecules is heavily influenced by the distance between the centers, distinction in oxidation-reduction potential and response of electron carriers to changes in charge on donor/acceptor molecules (Nicholls and Ferguson, 2002).

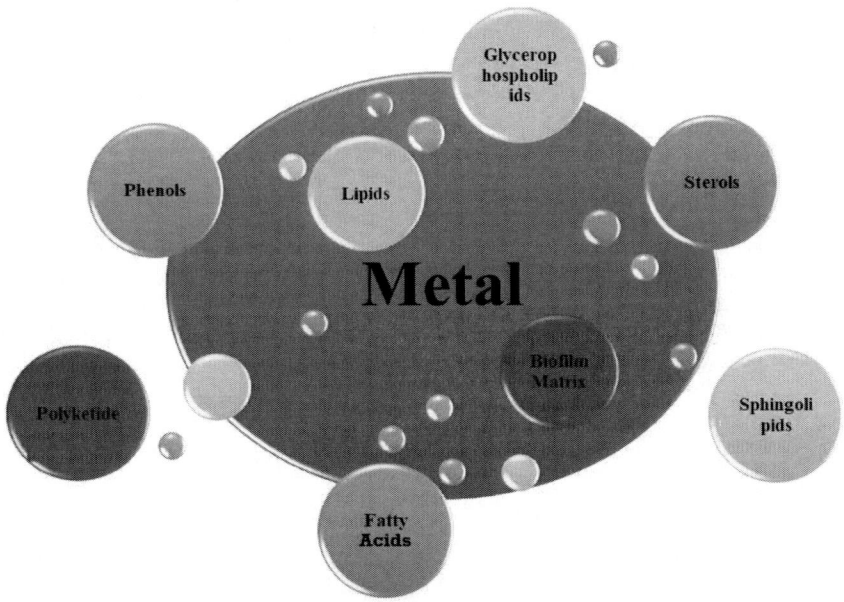

Figure 7. Role of lipids in corrosion induction.

Among the 2,353 putative metabolites, peptides and lipids were the numerically dominating compound groups, comprising 671 di-, tri-, and tetra-peptides and 590 lipids (sterol lipids, sphingolipids, prenols, polyketides, glycerophospholipids, and fatty acyls). While peptides were more numerous, lipids contributed ~4X as much as the peptides to the total metabolite abundance. A large group of putatively identified metabolites (247) were mapped to "housekeeping" processes including metabolism of amino acids (168) and carbohydrates (37). Secondary metabolism was represented by 110 putative compounds, while 735 putatively identified metabolites were not associated with any specific metabolic pathway in the

KEGG database. The richness of the HC metabolome was largely due to a wide variety of high-abundance lipids. Other putative metabolites preferentially present in the HC system were involved in biosynthesis and/or metabolism of steroid lipids, phenylpropanoids, porphyrins, arachidonic acid, unsaturated fatty acids, taurine, and isoprenoids. Further, iron ions were detected at high abundance in both HC samples, and they were four times more abundant in HC11 than in HC3. Technique to determine the lipids in corrosion process (Bonify et al., 2017; Beech and Gaylarde, 1999).

The functions and metabolism of fatty acids in biofilm have widely been studied by biologists, but it is still difficult to extract and exactly pinpoint the positions of these molecules within the cells using normal techniques. Fatty acids and their derivatives are necessary for the cell to perform and are associated with many alternative metabolic activities. Fatty acids that are produced by lipid reactions conjointly feed the citric acid cycle at the level of acetyl-CoA. The advanced method of extracting electrons from fat molecules for ATP production is named β-oxidation. Entry of fatty acyl-CoA into the mitochondrial matrix is prohibited unless the fatty acid is coupled via an ester linkage to carnitine that facilitates uptake of acyl molecules into the matrix, catalyzed by carnitine palmitoyltransferase1 (Cpt1). Carnitine is instantly exchanged with CoASH by Cpt_2, and acyl-CoA enters into β-oxidation once inside the matrix. It should be noted the yield of $1FADH_2$ and 1NADH along with an acetyl-CoA for every two carbons on the fatty acyl chain that enters into the cycle. Acetyl-CoA then enters the citric acid cycle where it is oxidized further (Mailloux, 2015). However, electron movement in mitochondria is much more difficult to characterize, given the completely different redox centers in mitochondrial enzymes.

Metabolic process complexes are separated by peptide chains with most carriers buried deep inside the lipid bilayer (Nicholls and Ferguson, 2002). Thus, electron transfer cannot be envisaged as an easy donation and acceptance of an electron between two completely different ions in an aqueous solution (Nicholls and Ferguson, 2002). Rather, electrons move between prosthetic groups and proceed via electron tunneling channels (Nicholls and Ferguson, 2002). Discussing the principles of electron transfer reactions in mitochondria is a vital consideration in the formation of O_2, which is presumably influenced by constant factors such as oxidation-reduction difference and response to donor/ acceptor molecules (Klinman, 2007). Peterson (1991) previously showed that UFAs facilitate electron transfer between iron centers including ferrous iron and ferric element cytochrome c. This author discovered that electron transfer is additionally increased during this process and has proposed a more physiological model of fatty acids related to proteins. Investigation increasingly assert that superoxide dismutase (SOD) increases electron transfer, while further examination continues to find whether or not free superoxide is involved in this electron transfer.

Both UFA and SOD are taking part in membrane redox systems, yet the mechanism of electron transfer remains largely unknown. Rate constants for photo induced electron transfer reactions of UFAs with a series of singlet excited states of oxidants in acetonitrile at 298 K were examined by Kitaguchi et al., (2006), and the ensuing electron transfer rate constants [k(et)] were evaluated in light of the free energy properties of electron transfer to determine the one electron oxidation potentials [E(ox)] of UFAs and therefore the intrinsic barrier to electron transfer. The k(et) values of linoleic acid with a series of oxidants are similar because the corresponding k(et) values of arachidonic acid, methyl linoleate and linolenic acid resulted in a similar E(ox) value for linoleic acid, methyl linoleate, linolenic acid and arachidonic acid (1.76 V vs. Singlet Excited States [SCE]). This is considerably less than that of monoUFA/oleic acid (2.03 V vs. SCE), as indicated by the k(et) values for oleic acid smaller than those of other UFA.

The unconventional ion of linoleic acid created photoinduced electron transfer from linoleic acid to the singlet excited state of 10-methyl acridinium ion, as well as that of 9,10-dihydroanthracene. It was detected by laser flash photolysis experiments. The apparent rate constant of deprotonation of the radical cation ion of linoleic acid was determined as 8.1 × $(10^3)^{s\,(-1)}$. No thermal electron transfer or proton-coupled electron transfer occurred from linoleic acid to a robust one electron oxidant, $Ru(bpy)_3^{(3+)}$ (bpy = 2,2'-bipyridine) or $Fe(bpy)_3^{(3+)}$ in the presence of oxygen, to the deprotonated radical produces the peroxyl radical. The above-mentioned proton transfer and electron transfer properties of UFAs give valuable mechanistic insight into lipoxygenases to clarify the proton-coupled electron transfer method in the catalytic function.

Sample Preparation

It is important that the sample chosen for analysis is representative of the lipids present in the biofilm, and that its properties are not altered prior to the analysis. Analysis of the types of lipids present in a sample usually requires that the lipid be available in a fairly pure form. For most, more rigorous extraction methods are needed, such as the solvent or non-solvent extraction methods. Once the lipids have been separated, they are often melted and then filtered or centrifuged to remove any extraneous matter. In addition, they are often dried to remove any residual moisture which might interfere with the analysis. As with any analytical procedure it is important not to alter the properties of the component being analyzed during the extraction process (Romanowicz et al., 2008).

Separation and Analysis by Chromatography

Chromatography is one of the most powerful analytical procedures for separating and analyzing the properties of lipids, especially when combined with techniques which can be used to identify the chemical structure of the peaks, e.g., mass spectrometry or NMR. A chromatographic analysis involves passing a mixture of the molecules to be separated through a column that contains a matrix capable of selectively retarding the flow of the molecules. Molecules in the mixture are separated because of their differing affinities for the matrix component in the column. The stronger the affinity between a specific molecule and the matrix, the more its movement is retarded and the slower it passes through the column. Thus, different molecules can be separated on the basis of the strength of their interaction with the matrix. After being separated by the column, the concentration of each of the molecules is determined as they pass through a suitable detector (e.g., UV-visible, fluorescence or flame ionization). Chromatography can be used to determine the complete profile of molecules present in a lipid. This information can be used to: calculate the amounts of saturated, unsaturated, polyunsaturated fat and cholesterol; the degree of lipid oxidation; the extent of heat or radiation damage; detect adulteration; determine the presence of antioxidants. Various forms of chromatography are available to analyze the lipids such as thin layer chromatography (TLC), gas chromatography (GC), and high-pressure liquid chromatography (HPLC) (Romanowicz et al., 2008).

Lipid Fractions by TLC

TLC is used mainly to separate and determine the concentration of different types of lipid groups such as triacylglycerols, diacylglycerols, monoacylglycerols, cholesterol, cholesterol oxides and phospholipids. A TLC plate is coated with a suitable absorbing material and placed into an appropriate solvent. A small amount of the lipid sample to be analyzed is spotted onto the TLC plate. With time the solvent moves up the plate due to capillary forces and separate different lipid fractions on the basis of their affinity for the absorbing material. At the end of the separation the plate is sprayed with a dye so as to make the spots visible. By comparing the distance that the spots move with standards of known composition it is possible to identify the lipids present. Spots can be scraped off and analyzed further using techniques, such as GC, NMR or mass spectrometry. This procedure is inexpensive and allows rapid analysis of lipids in fatty foods (Romanowicz et al., 2008).

Fatty Acid Methyl Esters by GC

Intact triacylglycerols and free fatty acids are not very volatile and are therefore difficult to analyze using GC (which requires that the lipids be highly non-polar and volatile). For this reason, lipids are usually derivatized prior to analysis to increase their volatility. Triacylglycerols are first saponified which breaks them down to glycerol and free fatty acids and are then methylated.

$$Triacylglycerol \xrightarrow{CH_2OH, N_2OH} Fatty\ acid\ methyl\ esters\ (FAMEs) + methylated\ glycerol$$

Saponification reduces the molecular weight and methylation reduces the polarity, both of which increase the volatility of the lipids. The concentration of different volatile fatty acid methyl esters (FAMEs) present in the sample is then analyzed using GC. The FAMES are dissolved in a suitable organic solvent that is then injected into a GC injection chamber. The sample is heated in the injection chamber to volatilize the FAMES and then carried into the separating column by a heated carrier gas. As the FAMES pass through the column, they are separated into a number of peaks based on differences in their molecular weights and polarities, which are quantified using a suitable detector. Determination of the total fatty acid profile allows one to calculate the type and concentration of fatty acids present in the original lipid sample (Eder, 1995).

Chemical Techniques

A number of chemical methods have been developed to provide information about the type of lipids present. These techniques are much cruder than chromatography techniques because they only give information about the average properties of the lipid components present, e.g., the average molecular weight, degree of unsaturation or amount of acids present. Nevertheless, they are simple to perform and do not require expensive apparatus, and so they are widely used in industry and research (Karn and Kumar, 2019).

Chapter 8

DNA or eDNA (Environmental DNA) in Corrosion Process

Abstract

Environmental DNA or eDNA is a nuclear or mitochondrial DNA that is detected in cellular or extracellular form and is released into the environment in free form from an organism. This environmental DNA is an effective tool used for the early detection of aquatic invasive species. Recently, environmental genomics has advanced greatly in a short period of time with demonstrable potential. Role of eDNA in corrosion process observed in specific metals such as silicon due to surface charge density and passivation conditions. Further, in the aquatic system, little information available about the factors that limits the persistence, detection and productions of eDNA.

Keywords: eDNA, DNA, organism, sea water, dissolved DNA

Introduction

DNA or eDNA

In organisms, deoxyribonucleic acid (DNA) is the hereditary material which codes for the biological instructions for maintaining and building them. All the organisms have the same chemical structure of DNA but differences lie in the order of DNA building blocks. Base pairs having unique sequences mainly repeating patterns that provide a means to identify populations, individuals and species. Environmental DNA (eDNA) is a mitochondrial or nuclear DNA which is released into the environment from an organism and may be detected in cellular or extracellular (dissolved DNA) form. Sources of eDNA such as shed skin and hair, mucous, carcasses, secreted faeces and gametes. On the basis of their environmental conditions, eDNA is distributed and diluted by water currents and other hydrological phenomena in the aquatic ecosystem but only lasts about 7-21 days (Dejean et al., 2011).

eDNA gets degraded by exposure to exo- and endonucleases, acidity, UVB radiation and heat (Karn et al., 2020).

eDNA: A Source For Detection of New Species

eDNA provides an attractive source for monitoring aquatic inventory with an aim to detect secretive, small, rare, and other species. Protocols using eDNA may allow for rapid, cost-effective, and standardized collection of data about species distribution and relative abundance. Consequently, biodiversity assessments can be significantly improved through detection of species using eDNA along with information with regard to habitat, status and distribution requirements for lesser-known or unknown species. Increasing evidence demonstrates improved species detection and catch-per-unit effort compared with electrofishing, snorkeling and other current field methods.

eDNA becomes an effective tool to early detect aquatic invasive species. For detection of invasive species, eDNA method may involve periodically collecting water samples and screening them for numerous invasive species simultaneously. For various invasive species, boat-ballast water such as mollusks becomes a source of introduction and might also be sampled. When limited surviving individuals recolonize the ecosystem then some of the intensive eradication programs for invasive species gets fail. The methods of eDNA provides a means of confirming the eradication of such invaders.

Later, oil and gas industry have recognized and valued the microbial diversity resources, becoming an integral component of the goals of sustainable development. The environmental genomics application has advanced greatly in a short period of time with demonstrable benefit potential.

eDNA in the Corrosion Process

Experimentally, DNA-induced corrosion behavior has been elucidated and is consistent with their porous element characterization via Fourier infrared prism and spectrometry coupling optical measurements, model proposed by Zhao et al., (2014). It has been found that this corrosion method and masks binding events improves by enhancing the concentration of either DNA targets or probes while the rate of corrosion is reduced via the passivation of the porous silicon surface by oxidation and salinization. A model has been devised in which corrosive porous silicon waveguides depends upon the surface charge density as well as passivation conditions of porous silicon structures Zhao et al., (2014). The strength of a negative charge related to DNA enhances once the surface coverage of the immobilized DNA molecules is enhanced (Figure 8).

The negative charge that is close to the surface of PSi attracts the majority of p-type PSi carriers to migrate to the neighborhood of DNA binding events that further reinforces the localized field of force close to the DNA-oxide interface. This electric field of force promotes the oxidation of Si pore wall and ends up in blue shift of Psi conductor reflection factor spectrum.

The mechanism behind the porous silicon corrosion is investigated by using another method (Lauw et al., 2010). At the surface of electrode as well as within the diffused layer, the double electrical layer having one polar element of the electrolyte becomes favorably accumulated while within the same region, the opposite polar element is exhausted.

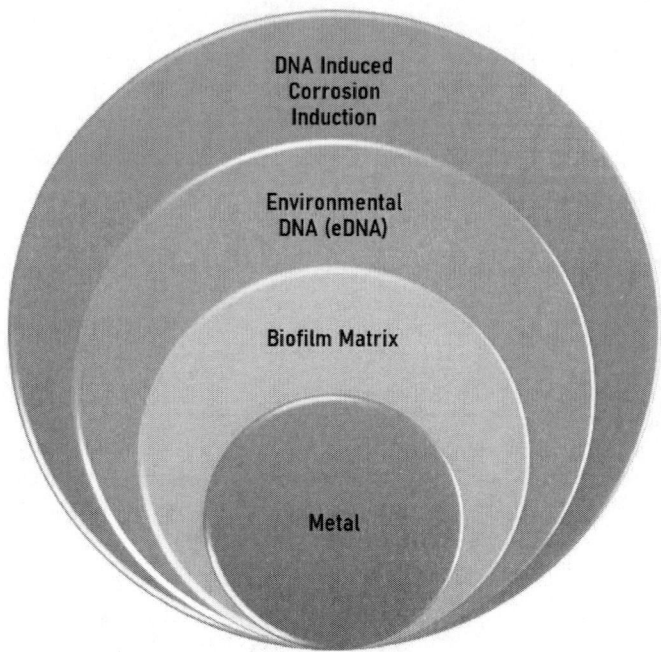

Figure 8. Role of DNA or eDNA in corrosion induction.

According to Liu et al., (2009) due to the increasing concentration of hydroxide around a hydrophobic nanotube, the double electrical layer accelerated the wet etching of SiO_2. Nevertheless, for the impact of corrosion, the formation of electrical double layers is unlikely to be the predominant reason as the hydrophilic as well as thermally oxidized surface of Psi is functionalized with the molecules of silane. Moreover, the negatively charged strands of DNA cannot be absorbed by the high effective hydroxide concentration. Consequently, it will accelerate SiO_2 etching.

Steinem et al., (2004) observed the optical thickness of a porous silicon matrix to reduce in close relation in to the presence of DNA. This thinning was interpreted as resulting from higher corrosion (oxidation hydrolysis) associated with DNA hybridization as well as negatively charged DNA accumulation close to the surface of Psi that increases the polarization of silicon bonds surface. It enables the nucleophilic attack by the molecules of water at exposed Si-H bonds. Biofilm matrix has another element i.e., extracellular DNA that is supposed to be a product of cell lysis. It plays a vital role in the structural stability as well as formation of biofilm whatsoever its origin. According to Watanabe et al., (1995) extracellular polymeric substances (nucleic acid) has been produced by the marine photosynthetic bacterium i.e., *Rhodovulum* sp. This nucleic acid plays a structural role due to the treatment of nucleolytic enzyme that resulted in deflocculation of the bacterium while polysaccharide-degrading enzymes like pectinase, trypsin and amylase had no effect. In several species, the extracellular DNA appears to be of diverse origins. Steinberger and Holden, (2005) demonstrated that genomic DNA as well as extracellular DNA found to be identical in *Pseudomonas aeruginosa* and *Pseudomonas putida* while Qin et al., (2007) demonstrated that due to autolysin AtlE, DNA originated from lysis of a population of attached microorganisms in *Pseudomonas epidermidis*.

Bockelmann et al., (2006) found that on the contrary, the extracellular DNA in aquatic bacteria of strain F8 was distinct from genomic DNA signifying a source diverse from cell lysis. On past and present biodiversity, DNA obtained from environmental samples such as water, sediments and ice (eDNA) represent a vital source of information. eDNA is essential for marine environment as well as industrial research. Behavior and origins of eDNA remains a very active research topic. eDNA can be deposited via saliva (Nichols et al., 2012), root cap cells and leaves (Trevors, 1996), skin flakes (Bunce et al., 2005), feathers (Taberlet and Bouvet, 1991), hair (Higuchi et al., 1988; Taberlet et al., 1993), pollen (Levy-Booth et al., 2007) regurgitation pellets (Taberlet and Fumagalli, 1996), urine (Valiere and Taberlet, 2000), faeces (Poinar, 1998), eggshells (Strausberger and Ashley, 2001), insect exuviae (Hofreiter et al., 2003) and in living prokaryotes through chromosomes such as in secretions and plasmids (Meier and Wackernagel, 2003).

The study of plants and microorganisms has shown that the dead and living cells can get lyses swiftly thereby releasing their DNA content (Nielsen et al., 2007). According to Crecchio and Stotzky, (1998), the survival of DNA is reinforced via binding to environmental compounds such as minerals, clay, different charged particles and larger organic molecules which shield adsorbed DNA from enzyme activity. Furthermore, this shielding from nucleases inhibits their ability to interact with the biota (Blum et al., 1997). According to Huang et al., (2014); Pietramellara et al., (2007), clay minerals such as montmorillonite will adsorb and absorb up to their own weight DNA and thus incorporate it onto to their large, charged area. Moreover, due to the negative charge on their surface, humic acids conjointly bind with DNA molecules and consequently extend survival of the DNA.

According to Lorenz and Wackernagel, (1987) the surface binding to sand is also possible that is markedly enhanced with the divalent cations concentrations like Mg^{2+} as well as Ca^{2+} readily forming sand-DNA bridges. It has been found that most of the studies of eDNA assumed that the age of the DNA molecule recovered is same as that of their matrix though, in certain conditions, molecules of DNA leach via the strata and contaminate lower layers (Haile et al., 2007). DNA can also leach through the sediments or frozen soil, currently, frozen soil has not been determined inevitably (Hebsgaard et al., 2009). Nevertheless, sediments leaching has been reported from both desert as well as temperate environments (Andersen et al., 2011; Jenkins et al., 2012), therefore, leaching possibility should always be taken into account and examined using independent methods (Haile et al., 2007). Even though the role of DNA in corrosion process has been treated in this segment, the analysis of eDNA has proved a powerful tool for the discovery of rich ocean biomes spanning the spectrum from the viruses via the bacteria to fish and mammals (Flaviani et al., 2017; Lacousière-Roussei et al., 2018). Figure 8 represents the role of DNA in corrosion process.

Even though the methods of eDNA measurement shows great potential for monitoring the inventory aquatic species, still there are plenty of details yet to be resolved. Further, in different types of aquatic systems, there is little information about factors which influence the lower limits of persistence, detection and production of DNA. These factors will probably vary amongst life stages and species. Further development and comparative testing of protocols are essential before the adoption of standard procedures for eDNA analysis and sampling.

Chapter 9

Technique to Identify Biomolecules or Biofilm in Corrosion

Abstract

Globally, corrosion is an issue which extensively affects many areas. It is a combination of physical, microbiological as well as chemical process that leads to material alteration. Biocorrosion is mainly induced by microorganisms which lead to the formation of biofilm. Biofilm causes the huge industrial economic loss. There are various techniques that are used to identify biofilms in corrosion such as confocal laser scanning microscopy, dry weight and total carbohydrates level, metal ion leaching analysis, microscopic analysis by scanning electron microscopy and energy-dispersive x-ray analysis and identification of biofilm bacteria by colony morphology, motility, oxidase (Difco), oxidation/fermentation of glucose and 16SrRNA Sequencing.

Keywords: biomolecules, biofilm, corrosion, technique aqueous environment

Introduction

Corrosion is a surface electrochemical phenomenon which is common to all base metals present in a humid environment. At the anodic sites, metal dissolution takes place at the separated sites with the acceptance of electrons in the cathodic pole. Under the acidic conditions at cathode, the most common reaction involves oxygen and it occurs at alkaline or neutral pH values. Nevertheless, protons can be the cathodic reactant that firstly gives rise to atom in the form of molecular hydrogen. Microorganism accumulates at the interfaces and forms polymicrobial aggregates like flocs, biofilms, films, sludge and mats. These microorganisms accumulate at the interfaces to form polymicrobial aggregates instead of living as pure cultures of dispersed single cells like sludge, biofilms, flocs, film and mat. There are various techniques to identify their role in corrosion process is mentioned in (Figure 9).

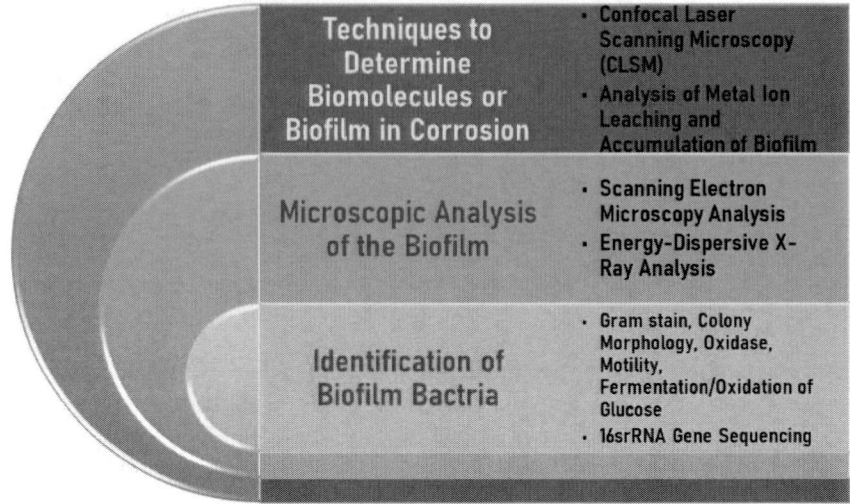

Figure 9. Techniques to determine biomolecules or biofilm in corrosion.

The matrix account for around 90% of the dry mass whereas microorganisms account for less than 10% in most of the biofilm structure. Matrix is the extracellular material that is formed by themselves by the organisms and the cells of biofilm get embedded in it. It involves accumulation of various kinds of biopolymers such as extracellular polymeric substances which provides the support for three-dimensional architecture of the biofilm and responsible for cohesion and surface adhesion in the biofilm.

There are various ways in which the cells of electrolytes become established such as within a single metallic structure or two dissimilar metals in electrical contact where different electrochemical potentials develop at different surface areas. This can be the result of surface imperfections, inclusions in the metal as well as presence of different concentration of cells.

Oxygen involves where the differential concentrations increase from the surface growths or deposits that limit the access of oxygen to the underlying metallic surface. Areas of low oxygen or other nutrient concentration are the anodic sites of metal dissolution. Due to the resistance of corrosion posed by surfaces such as stainless steel and aluminum due to the oxide passivation film formation where the cells of oxygen depletion have serious corrosive effect with these metals (Cragnolino and Tuovinen, 1984; Miller, 1981; Pope et al., 1984; Tiller, 1982; Tiller, 1983; Karn et al., 2020).

Confocal Laser Scanning Microscopy (CLSM)

Confocal laser scanning microscopy (CLSM) is the most essential tool for non-destructive *in-situ* components detection of the extracellular polymeric substances in the biofilms in combination with the fluorescent dyes (Lawrence et al., 2007). Exopolysaccharides are visualized according to their interaction with the exact target sugars by using fluorescently labelled lectins. In natural biofilms, such approaches revealed the complex arrangement and composition of EPS (Lawrence et al., 2007). Similarly, fluorescently labeled antibodies have been used against exopolysaccharides and this method is well established for use with pure cultures (Wrangstadh et al., 1990).

Non-specific binding patterns as well as clear-cut specific multi-labelling with both antibodies and lectins were reported by Neu and Lawrence, (1999). The use of Confocal Laser Scanning Microscopy based lectin-binding analysis in combination with Raman microscopy is the promising approach and this combination provides more in-depth insight into the extracellular polymeric substances' composition (Wagner et al., 2009). Nevertheless, substantial development is required for the distribution of spectra to individual lectin-stained clusters which is still a challenge. In biofilms, an approach to localizing the activity of enzymes is directs microscopic visualization by staining with fluorogenic substrates (Wingender and Jaeger, 2002).

In activated sludge flocs as well as in laboratory biofilms, the activity of phosphatase has been detected by using water-soluble substrate ELF-97 phosphate that yields an insoluble fluorescent precipitate upon cleavage by the enzyme. The spatial distribution of phosphatase activity has been studied in vertical sections as well as in whole flocs of biofilms. At the point of reaction, the reduction of tetrazolium salt 5-cyano-2,3-di-4-tolyl tetrazolium chloride (CTC) to CTC formazan crystals visualizes the extracellular redox activity. Specific dye for nucleic acid has been used in the detection of extracellular DNA. A 4-day-old culture of the gamma-proteobacterium strain F8 was grown on freshwater basal-medium agar that was isolated from the Saskatchewan River, Canada and afterwards stained with the dye SYTO9 to visualize the DNA. Next, between the strands of DNA, bacteria were visible as small rods apart from the DNA (Karn et al., 2020).

Determination of Dry Weight and Total Carbohydrate Levels

Biofilms can be removed from the pipe sections or the surfaces of steel which is 4 cm long of both grades. By using a sterile scalpel, biofilm is collected into a 10ml sterile double-distilled water. Further, the suspended biofilm formerly weighed and freeze-dried. Next, the samples were then analyzed for total carbohydrate by using Dubois, (1956) method using D-glucose as standard after resuspending in 1ml of distilled water.

Analysis of Metal Ion Leaching and Accumulation by Biofilm

Internal surfaces were washed with acidified water mainly (PH_3). Next, vigorously shaken the pipes for 4 mins. Further, inductively coupled plasma (ICP) spectrophotometer was used to analyze the acidified water for silicon, magnesium, chromium, aluminum, zinc, molybdenum, iron, phosphors, nickel and calcium ion simultaneously.

Microscopic Analysis of Biofilm

SEM and Energy-Dispersive X-Ray Analysis

SEM was used for the analysis of internal surface of the pipe of 2cm after air-drying the sputter-coated with gold. Sections of 1cm pipe coated with carbon by using carbon coater and commonly examined under a CamScan Series SEM with a Links EDX system.

Scanning Electron Microscopy (SEM) Analysis

After the formation of biofilm, SEM was used to examine the surface appearance of specimens. By using the pretreatments of carbon and stainless steel, SEM technique is used for observation of biofilm from the steel surface. At different time intervals, the pretreatment of biofilms was done by using ethanol at different concentrations (10%, 25%, 5%, 75% and 100 %). Next, treat them with glutaraldehyde and dried properly under vacuum and is used to observe the biofilms under SEM.

Weight Loss of Carbon Steel (Corrosion Rates)

The corrosion rate calculation was based on the weight loss analysis by measuring the mass of sample before and after the corrosion.

$$v = m_1 - m_2 \ AT_p$$

where, v is the rate of corrosion of carbon steel in mm y^{-1}, m_1 is the mass of coupon before the corrosion in g, m_2 is the coupon mass after the corrosion in g, A is the carbon steel surface area in mm$_2$, T is exposure of time in y and ρ is the metal density in g mm^{-3}.

All the measurements should be conducted in a sterile environment after 30 minutes of UV light sterilization to ensure the sterile environment.

Identification of Biofilm Bacteria

Isolated bacteria can be identified by biochemical characterization such as catalase (DIFCO), gram stain, growth at different temperatures such as 37°C, 41°C and 45°C, colony morphology, oxidase (DIFCO), fermentation/ oxidation of glucose, motility (hanging drop method) and transmission electron microscopy observation for the presence of polar flagella (Karn et al., 2010, 2011).

Further gram-negative bacteria identification can be carried out on the basis of their color, catalase (Difco), gram stain. Next, culture maintained on R3A medium at 4°C in a pure state (Reasoner and Geldreich, 1985). LeChevallier, (1980) stated that gram-negative as well as gram-positive bacteria could be closely identified with an identification method formulated by the verification of identity of 70-80Y0 of predominant bacteria accepted by the National Collection of Industrial and Marine Bacteria (NCIMB, Aberdeen). Further, identification of the organism can be done by 16S rRNA Gene Sequencing.

Kapley et al., (2001) suggested genomic DNA could be extracted from the overnight grown culture. Purified genomic DNA served as a template. Further, Gene Amp 2700 PCR systems (Applied Biosystems, USA) could be used to amplify 16S rRNA gene with forward primer: 5'-AGAGTT-TGATCCTGGCTCAG-3' and reverse primer: 5'-ACGGGCGGTGTGTTC-3' (Weisburg et al., 1991). In PCR, the reaction mixture contains PCR buffer, 200 mM of dNTPs, 1.5 mM MgCl$_2$, 0.1 mM of each primer and 2.5

units of Taq DNA polymerase (Invitrogen, USA) and make up to a final volume of 100 sterile MQ water.

Furthermore, PCR is performed with an initial denaturation at 92°C for 2 min followed by 36 cycles of 92°C for 1 min, 48°C for 30 s and 72°C for 2 min and a final extension of 72°C for 6 min. By using QIA gel extraction kit, the amplification product is gel purified (Qiagen, USA) it gets ligated into the pGEM-T easy vector. Next, by using heat shock method and provides calcium chloride treatment, the ligated plasmid transforms into the cells of *E. coli*. By using the chain termination method, the partial sequence was generated after the screening of positive clones by using an applied biosystems automated DNA sequence. In GenBank[1] the sequence further compares against the available DNA sequences from the type strains by using BLASTN sequence match tool. Afterwards, the sequences align by using MultAlin[2] program and further alignment gets corrected manually and later phylogenetic tree can be constructed.

---

[1] http://www.ncbi.nlm.nih.gov/.
[2] http://bioinfo.genotoul.fr/multalin/multalin.html.

Chapter 10

Omics Approach in Biocorrosion

Abstract

For the biological sciences, developing new discipline has vital possible implications. Metabolomics, genomics, proteomics, transcriptomics is an emerging field of science. Worldwide, biochemical events are studied by metabolomics, genomics and proteomics by using the assays of small molecules in biological fluids, biofilm, cells, organs as well as tissues followed by the informatics techniques to define the identities of metabolomics. The understanding of their biological mechanisms enhances by the studies of metabolomics, genomics, proteomics, transcriptomics. This chapter deals with several conceptual approaches as well as the experimental development for the genomics and proteomics that applied for the analysis of molecular connectivity of biofilm as well as metabolic foot printing, and elucidation of mineralization process.

Keywords: metabolomics, proteins, transcriptome, applied biology, bioremediation, genomics, proteomics

Omics Approach

Metabolomics, genomics, proteomics as well as transcriptomics has high impact on the conducted research in various application areas. At the metabolite level, metabolomics accompanies proteomics, genomics and transcriptomics (Fiehn, 2002). Apparently, the metabolome is determined by the coding capacity of exon the genome, and it is influenced by changes in the transcriptome and proteome but also by the direct or post-translational impact of the chemical and nutritional environment of a cell on enzymatic and transport activities. In comparison to proteomics and transcriptomics, metabolomic analyses represent the level most closely related to the physiology of an organism, organ, or cell (Roessner et al., 2003; Desbrosses et al., 2005).

Metabolomics is well-thought-out to provide a direct functional readout of the physiological state of an organism (Gieger et al., 2008). To examine the metabolites in various tissues, fluids and organisms, a range of analytical technologies has been employed. Mass spectrometry when coupled with different chromatographic separation techniques like nuclear magnetic resonance or gas or liquid chromatography is the major tool to examine the large number of metabolites concurrently. Even though the technology is extremely sensitive and sophisticated, still there are few bottlenecks in metabolomics, genomics, proteomics as well as transcriptomics. There is no single technology that is accessible to date analyzes the complete metabolome, proteome, transcriptome and genome. Consequently, established the number of complementary approaches for identification, detection, extraction and quantification of many possible metabolites (Villas-Boas et al., 2007).

In the metabolic reaction network, the level of metabolite led to the bottleneck's identification. Additionally, in *in-vivo* reactions, metabolomics created reactants as well as products that directly connect with the cellular metabolism of phenotype rather than proteome as well as transcriptome (Oliver et al., 1998). Metabolomics consequently complements the proteomics, genomics, fluxomics and transcriptomics data as well as enable the efforts of system biology and metabolic engineering towards the designing of cell factories and superior biocatalysts. Small molecules of biomarkers have been identified potentially by metabolomics which correlates and applies in research and delivers early information for various forms of cancer, environmental or genetic variation, drug discovery, insulin resistance, diseases as well as drug toxicity and biofilm connectivity.

Fast adoption of omics approach such as transcriptomics, metabolomics, genomics and proteomics to analyze biological samples generates tera-to peta-byte sized data files on regular basis. They are diversely called as trans-omics, integrated omics, poly-omics, pan-omics, multi-omics or usually called as omics. They are important tool in meeting challenges like differences in biomolecule identification, data storage and handling, normalization, sharing and data archiving, data dimensionality reduction, data cleaning, statistical validation, biological contextualization. The final goal is to achieve near the holistic realization of systems biology towards the understanding of biological questions. Although there is decrease in the processing as well as costs and also datasets generated is of much higher quality various scientists face challenges that are in this interdisciplinary domain of bioinformatics.

It is difficult and many times quite challenging to comprehend these. Pavlopoulos et al., (2015) described that further supplementary data enhances the complexity from DNA/protein, microRNA/gene, and protein/RNA as well as protein/protein interactions. To visualize and analyze the proteomics, software, tools, databases, metabolomics and resources data gets reviewed (Ovel et al., 2015; Misraand van der Hooft, 2016; Misra et al., 2017; Misra, 2018).

Although the importance of omics data integration has been realized for a broad range of research areas such as phenotype-genotype interactions (Ritchie et al., 2015), nutrition and food science (Kato et al., 2011), analysis of microbiomes (Muller et al., 2014), systems microbiology (Fondi and Lio, 2015), successful implementation of more than two omics datasets is very rare. Gehlenborg et al., (2010) produced a useful comprehensive compendium for the omics data visualization for systems biology using data from mass spectrometry, nuclear magnetic resonance, microarrays, RNA deep sequencing and protein interactions. Considerable progress has been made for development of additional tools and approaches for the integrated omics analysis.

In the late 1970, the emergence of recombinant DNA technology led direct overview of specific product pathway of interest by using a specific gene like insulin production by *Saccharomyces cerevisiae* as well as *Escherichia coli* correspondingly and the flux enhancement to the products that produced naturally by the organisms (Nielsen, 2001). Various products that derived from the microorganisms are not directly linked with the specific growth rate. Nevertheless, the concentrations as well as the production rate are linked indirectly to the primary metabolisms as ATP, precursors as well as co-factors in the form of NAD (H) and NADP (H) has been provided by the primary central carbon metabolism. Consequently, the engineering of superior bacterial cells for the improvement of microbial products production by using the laboratory strains to date requires firm understanding of primary cellular metabolism and its regulation *in vivo* (Bailey, 1991).

Metabolomics: An Approach

In metabolomics, there are four conceptual approaches such as metabolic fingerprinting, metabolite profiling, target analysis and metabolomics (Fiehn, 2002). For many decades, target analysis applied and comprises the quantification as well as the determination of small set of known metabolites by using a particular technique for compounds of interests.

On the other hand, metabolite profiling aims is to analyze the larger sets of compounds both identified as well as unknown with respect to their chemical nature. For various biological systems, this approach has been applied by using the gas chromatography-mass spectrometry for plasma, plants, urine and microbial samples. Complementary analytical methodologies employ metabolomics, genomics, proteomics as well as transcriptomics such as liquid chromatography-mass spectrometry, nuclear magnetic resonance as well as gas-chromatography-mass spectrometry for the determination as well as quantification of as many metabolites as possible as well as identified unknown compounds.

Metabolic Fingerprinting

Here a metabolic signature or mass profile of the sample of interest is generated and then compared in a large sample population to screen for differences between the samples. When signals that can significantly discriminate between samples are detected, the metabolites are identified and the biological relevance of that compound can be elucidated, greatly reducing the analysis time. Since metabolites are so closely linked to the phenotype of an organism, metabolomics can be used for a large range of applications, including phenotyping of genetically modified plants and substantial equivalence testing, determination of gene function, and monitoring responses to biotic and abiotic stress. Metabolomics can therefore be seen as bridging the gap between genotype and phenotype, providing a more comprehensive view of how cells function, as well as identifying novel or striking changes in specific metabolites (Fiehn, 2002). Analysis and data mining of metabolomic data sets and their metadata can also lead to new hypotheses and new targets for biotechnology.

Metabolomics in Biocorrosion

Metabolome analysis is a powerful approach to discover novel metabolic pathways and characterize metabolic networks. Physiological responses are represented by metabolomics, genomics, proteomics data for the genetic, environmental as well as developmental changes (Kell et al., 2005).

To elucidate the mineralization pathway of metals in biofilm, modern high-end techniques such as isotope distribution analysis of metabolites as well as molecular connectivity analysis by using Ultra-high mass accuracy techniques, Nuclear magnetic resonance, Spectroscopy, Mass Spectrometry based assay, can be utilized using Metabolic foot printing analysis. Lately, extracellular metabolite profiles measurement is known as a powerful tool in metabolic foot printing analysis for biofilm, microbial mutant classification for functional genomics and is focused on the biochemical as well as chemical alterations measurement provoked by an organism in any environment such as laboratory culture medium (Villas-Boas et al., 2006). To the extracellular medium, a living cell excretes the metabolites as well as secretes enzymes and these metabolites and enzymes interact as well as modify the medium components, resulting in highly specific metabolic profiles to genetic or species backgrounds (Kell et al., 2005).

Analysis of Metabolic Foot Printing

Metabolic foot printing can be a very useful approach to evaluate the process of mineralization of xenobiotics in the laboratory or in the environment. For the analysis of metabolic foot printing, analytical techniques aim to be an unbiased and non-targeted analysis of low molecular mass compounds in a broad hypothesis-driven approach. Consequently, it is a highly recommended approach to confirm that whether a polluting compound is degraded completely to water and CO_2 or the process of biodegradation resulting in the accumulation of recalcitrant and hazardous catabolic products which are bottlenecking the process of mineralization.

Consequently, metabolomics compliments proteomics, genomics, flux omics as well as transcriptomics data and enable system biology as well as metabolic engineering efforts towards the cell factories as well as superior biocatalysts designing (Figure 10). Quantitative understanding of microbial metabolism and its *in vivo* regulation required knowledge of both extracellular and intracellular metabolites. Traditionally, this knowledge is acquired through fast sampling, instant arrest of metabolic activity and deactivation of endogenous enzymatic activity, metabolite extraction and subsequent quantification of intracellular reactants (metabolites). Extracellular metabolites are quantified in the cell free supernatant obtained either by filtration or centrifugation at low temperatures.

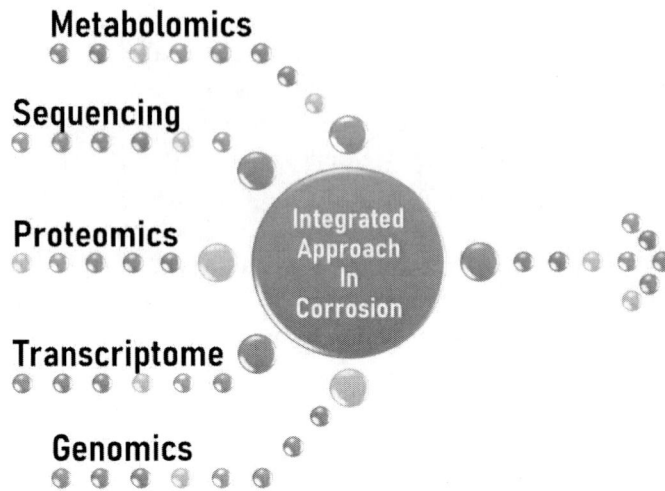

Figure 10. Role of omics in the diagnosis of corrosion mechanism.

The ongoing question towards understanding *in-vivo* regulation of microbial metabolic networks have been the primary fuel for the recent rapid developments in metabolomics i.e., quantification of the total complement of metabolites inside (endometabolome) and outside (exometabolome) a cell in different environments, growth conditions or genetic perturbations. However, there is hardly any method available for the determination of low molecular weight metabolite, hence; there is a lot of grey area in this field. For the determination of single analyte, it is necessary to depend upon multiple methods. Additionally, single method that separates all the metabolites appears to be unlikely and gives wide diversity in physical as well as chemical properties inherent to the metabolites constituting the bacterial metabolome. Consequently, to develop the techniques, it appears more practical and dedicated as well as targeting the classes of metabolites.

Metabolomics, genomics, proteomics and transcriptomics development database contains accurate measured concentration of metabolite under the given sets of standard culture conditions that serves as a guide and position metabolomics as an important part of the bacterial research and technology. Omics approach is a young field in the growth phase. In this emerging area of science, the field of mineralization has gained more from the advances. By using various methods, the data from protein, genetics, metabolite and transcripts level from the organisms has been obtained and when this data is compared with the other organisms, it becomes valuable information in the system biology.

References

Chapter 1

Ahmadkhaniha, D., Järvenpää, A., Jaskari, M., Sohi, M. H., Zarei-Hanzaki, A., Fedel, M., Deflorian, F., Karjalainen, L. P., Mech. J. (2016). *Behav. Biomed. Mater.*, 61: 360-370.

Aktas, D. F., Sorrell, K. R., Duncan, K. E., Wawrik, B. Callaghan, A. V., Suflita, J. M. (2017). *Int. Biodeterior. Biodegrad.*, 118: 45-56.

Angell, P. and Chamberlain, A. H. L. (1991). The role of extracellular products in copper colonization. *Int. Biodeterior.*, 27: 135- 143.

Araujo-Jorge, T. C., Coutinho, C. M. L. and Aguiar, L. E. V. (1992). Sulphate-reducing bacteria associated with biocorrosion: a review. *Memórias do Instituto Oswaldo Cruz.* 87(3): 329- 337.

Arens, P., Tuschewitzki, G. J., Wollmann, M., Follner, H. and Jacobi, H. (1995). Indicators for microbiologically induced corrosion of copper pipes in a cold-water plumbing system. *Int. J. Hyg. Environ. Med.*, 196: 444- 454.

Ashassi-Sorkhabi, H., Moradi-Haghighi, M. and Zarrini, G. (2012). *Mater. Sci. Eng.*, C32: 303- 309.

Beech, I. B. (2004). Corrosion of technical materials in the presence of biofilms–current understanding and state-of-the art methods of study. *Int. Biodeter. Biodegr.*, 53: 177- 183.

Beech, I. B., and Cheung, C. W. S. (1995). Interactions of exopolymers produced by sulphate-reducing bacteria with metal ions. *International Biodeterioration & Biodegradation.*, 35(1-3): 59- 72.

Beech, I. B., Cheung, C. W. S., Johnson, D. B., & Smith, J. R. (1996). Comparative studies of bacterial biofilms on steel surfaces using atomic force microscopy and environmental Scanning Electron Microscopy. *Biofouling.*, 10(1): 65- 77.

Beech, I. B. and J. Sunner, (2004). Biocorrosion: towards understanding interactions between biofilms and metals. *Curr. Opin. Biotechnol.*, 15: 181- 186.

Beech, I. B., Sunny Cheung, C. W., Patrick Chan, C. S., Hill, M. A., Franco, R., & Lino, A. R. (1994). Study of parameters implicated in the biodeterioration of mild steel in the presence of different species of sulphate-reducing bacteria. *International Biodeterioration and Biodegradation.*, 34(3-4): 289- 303.

Beech, I. B., Sunny Cheung, C. W., Patrick Chan, C. S., Hill, M. A., Franco, R., & Lino, A. R. (1998). Direct involvement of an extracellular complex produced by a marine sulfate reducing bacterium in deterioration of steel. *Geomicrobiology Journal.*, 15(2): 121- 134.

Berndt, M. L. (2011). *Constr. Build. Mater.*, 25: 3893- 3902.

Bhat, S., Sharma, V. K., Thomas, S., Anto, P. F., Singh, S. K. (2011). *Mater. Perform.* 50: 50- 53.

Boopathy, R., Robichaux, M., LaFont, D., Howell, M. (2002). Activity of sulfate-reducing bacteria in human periodontal pocket. *Can. J. Microbiol.*, 48: 1099-1103.

References

Borenshtein, S. W. (1994). Microbiologically Influenced Corrosion Handbook. *Woodhead Publishing Limited, Cambridge, UK.*

Brooks, W. (2013). Corrosion, *NACE International, Orlando, Florida.*, 02525.

Castaneda, H. and Benetton, X. D. (2008). SRB-biofilm influence in active corrosion sites formed at the steel-electrolyte interface when exposed to artificial seawater conditions. *Corrosion Science.*, 50(4): 1169- 1183.

Chandrasatheesh, C., Jayapriya, J., George, R. P., KamachiMudali, U. (2014). *Eng. Fail. Anal.*, 42: 133- 142.

Ching, T. H., Yoza, B. A. Wang, R., Masutani, S., Donachie, S., Hihara, L. and Li, Q. X. (2016). *Int. Biodeterior. Biodegrad.*, 108: 122- 126.

Coetser, S. and Cloete, T. (2005). Biofouling and biocorrosion in industrial water systems. *Critical Reviews in Microbiology.*, 31(4): 213- 232.

Cord-Ruwisch, R., Kleinitz W. and Widdel, F. (1987). Sulfate-reducing bacteria and their activities in oil production. *Journal of Petroleum Technology.*, 39(1): 97- 106.

Costello, J. A. (1974). Cathodic depolarization by sulphate-reducing bacteria. *South African Journal of Science.*, 70(7): 202- 204.

Critchley, M. M. and Fallowfield, H. J. (2001). The effect of distribution system bacterial biofilms on copper concentrations in drinking water. *Water Sci. Technol.*, 247-252.

Critchley, M. M., Pasetto, R. and O'Halloran, R. J. (2004). Microbiological influences in blue water copper corrosion. *J. Appl. Microbiol.*, 97: 590- 597.

Crolet, J. (1992). Biocorrosion: pH Regulation by Sulphate-Reduction Bacteria. *Mater. Tech. (Paris).* 80(9-10): 71- 77.

Dai, X., Wang, H., Ju, L. K., Cheng, G., Cong, H. and B. Z. (2016). Newby. *Int. Biodeterior. Biodegrad.* 115: 1- 10.

Dom alicki, P., E. Lunarska and J. Birn, (2007). Effect of cathodic polarization and sulfate-reducing bacteria on mechanical properties of different steels in synthetic sea water. *Materials and Corrosion.*, 58(6): 413- 421.

Edyvean, R., Berrson, J., Thomas, C. J., I., Beech, B., Videla, H. A. (1998). Biological influences on hydrogen effects in steel in seawater. *Materials Performance.*, 37(4): 40- 44.

Fang, H. H. P., Xu, L. C. and Chan, K. Y. (2000). Influence of Cr3+ on microbial cluster formation in biofilm and on steel corrosion. *Biotechnology Letters.*, 22(9): 801- 805.

Fang, H. H. P., Xu, L. C. and Chan, K. Y. (2002). Effects of toxic metals and chemicals on biofilm and biocorrosion. *Water Research.*, 36(19): 4709- 4716.

Flemming, H. C. (1996). Biofouling and microbiologically influenced corrosion (MIC) an economical and technical overview. In: Heitz, E., Sand, W., Flemming, H. C. (Eds.), *Microbial Deterioration of Materials.* Springer, Heidelberg.

Fontana, M. G. (1986). *Corrosion Engineering*, third ed., McGraw-Hill, New York.

Ford, T. and Mitchell, R. (1990). The ecology and microbial corrosion. *Advances in Microbial Ecology.*, 11: 231- 262.

Gu, T. and Xu, D. (2010). Demystifying MIC Mechanisms. *Corrosion.* 2010.

Gu, T. and Xu, D. (2013). Why are some microbes corrosive and some not? Corrosion/2013, Paper No. C2013-0002336, *NACE International, Houston, TX.*

References

Gu, T., Zhao, K. and Nesic, S. (2009). A Practical Mechanistic Model for MIC Based on a Biocatalytic Cathodic Sulfate Reduction Theory. National Association of Corrosion Engineers, P. O. Box 218340 Houston TX 77084 USA.

Hamilton, W., (1985). Sulphate-reducing bacteria and anaerobic corrosion. *Annual Reviews in Microbiology.*, 39(1): 195- 217.

Harbulakova, V. O., Estokova, A., Stevulova, N., Luptáková, A. and Foraiova, K. (2013). *Procedia Eng.*, 65: 346- 351.

Heitz, E., Flemming, H. C., Sand W. (Eds.). (1996). *Microbially Influenced Corrosion of Materials,* Springer-Verlag GmbH & Co KG, Berlin, Germany.

Heyer, A., D'Souza, F., Morales, C. F. L., Ferrari, G., Mol, J. M. C., de Wit, J. H. W. (2013). *Ocean Eng.*, 70: 188- 200.

Hiibel, S. R., Pereyra, L. P., Inman, L. Y., Tischer, A., Reisman, D. J., Reardon, K. F. and Pruden, A. (2008). Microbial community analysis of two field-scale sulfate-reducing bioreactors treating mine drainage. *Environ. Microbiol.,* 10: 2087- 2097.

Hilbert, L. R., Hemmingsen, T., Nielsen, L. V., & Richter, S. (2005). When Can Electrochemical Techniques Give Reliable Corrosion Rates on Carbon Steel in Sulfide Media? *Corrosion.*

Hinkson, D., Wheeler, C. and Oney, C. (2013). Corrosion/2013. *NACE International, Orlando, Florida.*, 02276.

Hou, B., Li, X., Ma, X., Du, C., Zhang, D., Zheng, M., Xu, W., Lu, D., Ma, F. (2017). *Npj Mater. Degrad.*, 1: 4.

Ilhan-Sungur, E., Cansever, N., Cotuk, A. (2007). Microbial corrosion of galvanized steel by a freshwater strain of sulphate reducing bacteria (*Desulfovibrio* sp.), *Corr. Sci.* 49: 1097- 1109.

Iverson, P. (2001). Research on the mechanisms of anaerobic corrosion. *International Biodeterioration & Biodegradation.*, 47(2): 63- 70.

Iverson, W. and G. Olson. (1983). Anaerobic corrosion by sulfate-reducing bacteria due to highly reactive volatile phosphorus compound. *Microbial Corrosion.*, 46- 53.

Iverson, W. P. (1968). Corrosion of iron and formation of iron phosphide by *DesulfovibrioDesulfuricans. Nature,* 217: 1265- 1267.

Jack, R. F., Ringelberg, D. B., White, D. C. (1992). Differential corrosion rates of carbon steel by combinations of *Bacillus* sp., Hafnia alvei, and *Desulfovibrio*gigas established by phospholipid analysis of electrode biofilm. *Corros. Sci.* 33: 1843- 1853.

Jacobson, G. A. (2007). *Mater. Perform.* 46: 26- 34.

Javaherdashti, R. (2016). Microbiologically Influenced Corrosion: *An Engineering Insight,* Springer, London.

Javaherdashti, R. A review of some characteristics of MIC caused by sulfate-reducing bacteria: past, present and future. (1999). *Anti-Corros. Methods Mater.*, 46: 173-180.

Jia, R., Yang, D., Abd Rahman, H. B. and Gu, T. (2017). *Int. Biodeterior. Biodegrad.,* 125: 116- 124.

Kakooei, S., Ismail, M. C. and Ariwahjoedi, B. (2012). Mechanisms of Microbiologically Influenced Corrosion: A Review. *World Applied Sciences Journal.* 17(4): 524- 531.

King, R. A. and Miller, J. D. A. (1971). Corrosion by sulphate-reducing bacteria. *Nature.* 233: 491- 492.

King, R. A., Miller, J. D. A. and JS, S. (1973). Corrosion of mild steel by iron sulfides. *British Corrosion Journal.*, 8: 137- 142.

Kip, N. and Veen, J. A. V. (2015). The dual role of microbes in corrosion. *ISME J.*, 9: 542- 551.

Koch, G. H., Brongers, M. P. H., Thompson, N. G., Virmani, Y. P. and Payer, J. H. (2018). Corrosion Cost and Preventive Strategies in the United States., http://trid.trb.org/view.aspx?id=707382, April 6, 2014.

Koch, J. H., Brongers, M. P. H., Thompson, N. G., Virmani, Y. P. and Payer, J. H. (2002). Corrosion Cost and Preventive Strategies in the United States, Federal Highway Administration, Washington, DC, Report No. FHWA-RD-01-156.

Kuehr, V. W., H, C. A. and Vlugt, I. S. V. D. (1934). The graphitization of cast iron as an electrobiochemical process in anaerobic soil. *Water.*, 18: 147- 165.

Kumar, C. and S. Anand., (1998). Significance of microbial biofilms in food industry: a review. *International Journal of Food Microbiology.*, 42(9): 27.

Lee, W. and W. Characklis. (1993). Corrosion of mild steel under anaerobic Biofilm. *Corrosion*, 49(03).

Lee, W., Lewandowski, Z., Nielsen, P. H., & Hamilton, W. A. (1995). Role of sulfate-reducing bacteria in corrosion of mild steel: a review. *Biofouling.*, 8(3): 165- 194.

Li, H., Yang, C., Zhou, E., Yang, C., Feng, H., Jiang, Z., Xu, D., Gu, T., Yang, K. and Mater. (2017). *J. Sci. Technol.*, 33: 1596- 1603.

Li, H., Zhou, E., Zhang, D., Xu, D., Xia, J., Yang, C., Feng, H., Jiang, Z., Li, X., Gu, T. and Yang, K. (2016). *Sci. Rep.*, 6: 20190.

Li, P., Zhao, Y., Liu, Y., Zhao, Y., Xu, D., Yang, C., Zhang, T., Gu, T. and Yang, K. (2017). *Mater. Sci. Technol.*, 33: 723- 727.

Li, X., Zhang, D., Liu, Z., Li, Z., Du, C. and Dong, C. (2015). *Nature.*, 527: 441- 442.

Li, Y., Jia, R., Al-Mahamedh, H. H., Xu, D. and Gu, T. (2016). *Front. Microbiol.* 7: 896.

Liamleam, W. and Annachhatre, A. P. (2007). Electron donors for biological sulfate reduction. *Biotechnol. Adv.*, 25: 452- 463.

Little, B. J., Pope, R. and Ray, R. (2000). Relationship between corrosion and the biological sulfur cycle: a review. *Corrosion.*, 56(04).

Little, B. Wagner, P., Hart, K., Ray, R., Lavoie, D., Nealson K. & Aguilar C. (1998). The role of biomineralization in microbiologically influenced corrosion. *Biodegradation.*, 9(1): 1- 10.

Lopez, M. A., Diaz de la Serna, F. J. Z., Jan-Roblero, J., Romero, J. M., Hernandez-Rodriguez, C. (2006). Phylogenetic analysis of a biofilm bacterial population in a water pipeline in the Gulf of Mexico. *FEMS Microbiol. Ecol.*, 58: 145-154.

Marcus, P. (2002). Corrosion mechanisms in theory and practice: CRC.

Maria, L., Carvalho, Doma, J., Sztyler, M, Beech, I. and Cristiani, P. (2014). The study of marine corrosion of copper alloys in chlorinated condenser cooling circuits: The role of microbiological components. *Bioelectrochemistry.* 97: 2- 6.

Maruthamuthu, S., Dhandapani, P., Kamalasekaran, S., Siddarth, A. S., Manoharan, S. P., Muthuraman, K. and Narayanan, G. (2013). *Eng. Fail. Anal.*, 33: 315- 326.

Mehana, M. (2009). Mécanismes de transfert direct en corrosion microbienne des aciers: Application à Geobactersulfurreducens et à l'hdrogénase de Clostridium acetobutylicum [*Mechanisms of direct transfer in microbial corrosion of steels:*

References

Application to *Geobactersulfurreducens* and the hydrogenase of *Clostridium acetobutylicum*], (Thesis Université de Toulouse) 2009.

Miranda, E., Bethencourt, M., Botana, F. J., Cano, M. J., Sánchez-Amaya, M. J. (2006). Corzo, A., García de Lomas, J., Fardeau, M. J., Ollivier, B. Biocorrosion of carbon steel alloys by an hydrogenotrophic sulphate-reducing bacterium *Desulfovibriocapillatus* isolated from a Mexican oil field separator. *Corr. Sci.*, 48: 2417- 2431.

Momba, M. N. B., Kfir, R., Venter, S. N., Cloete, T, E. 2000. Overview of biofilm formation in distribution systems and its impact on the deterioration of water quality.

Obuekwe, C., Westlake, D. and Plambeck, J. (1981). Corrosion of Mild Steel in Cultures of Ferric Iron. Reducing Bacterium Isolated from Crude Oil. II.-- Mechanism of Anodic Depolarization. *Corrosion.*, 37(11): 632- 637.

Ocando, L. Matilde de Romero, Ennery Leon, Laura Atencio, Orlando Perez, and Oladis de Rincon (2007). Evaluation of pH and H_2S on biofilms generated by sulfate-reducing bacteria: influence of ferrous ions. *Corrosion.*

Qian, H., Li, M., Li, Z., Lou, Y., Huang, L., Zhang, D., Xu, D., Du, C., Lu, L. and Gao, J. (2017). *Mater. Sci. Eng.C.*, 80: 566- 577.

Romero, M. (2005). The Mechanism of SRB Action in MIC, Based on Sulfide Corrosion and Iron Sulfide Corrosion Products. *Corrosion*, (2005).

San, N. O., Nazır, H. and DönmezCorros, G. (2012). *Sci.*, 64: 198- 203.

Santana Rodríguez, J. J., Santana Hernández, F. J., González González, J. E. (2006). Comparative study of the behaviour of 304 SS in a natural seawater hopper, in sterile media and with SRB using electrochemical techniques and SEM. *Corr. Sci.* 48, 1265- 1278.

Sheng, X., Ting, Y. P. and Pehkonen S. O. (2007). The influence of sulfate-reducing bacteria biofilm on the corrosion of stainless steel AISI 316. *Corr. Sci.* 49: 2159- 2176.

Sowards, J. W., Williamson, C. H. D. and Weeks, T. S., McColskey, J. D., Spear, J. R. (2014). *Corros. Sci.* 79: 128- 138.

Sowards, W. and Mansfield, E. (2014). *Corros. Sci.*, 87: 460- 471.

Tan, J. L., Goh, P. C., Blackwood, D. J. (2017). *Corros. Sci.*, 119: 102-111.

Teng, F., Guan, Y. T., Zhu, W. P. (2008). *Corros. Sci.*, 50: 2816- 2823.

Thauer, R. K., Stackebrandt, E. and Hamilton, W. A. (2007). Energy metabolism and phylogenetic diversity of sulphate-reducing bacteria, in: L.L. Barton, W.A. Hamilton (Eds.), Sulphate-reducing Bacteria: Environmental and Engineered Systems, Cambridge University Press, New York, 1- 37.

Veronika, K. B. (2008). Knowledge about metals in the first century. *Korroz, Figy.*, 48: 133-137.

Videla, H. A. (1996). *Manual of Biocorrosion*, CRC Press, Boca Raton, FL, USA.

Videla, H. A. (2000). An overview of mechanisms by which sulphate-reducing bacteria influence corrosion of steel in marine environments. *Biofouling.*, 15(1-3): 37-47.

Videla, H. A., Edyvean, G. and Herrera, L. K. (2005). An updated overview of SRB induced corrosion and protection of carbon steel. *Corrosion.*

Videla, H. A. and Herrera, L. K. (2005). Microbiologically influenced corrosion: looking to the future. *Int. Microbiol.*, 8: 169- 180.

von WolzogenKuehr, C. A. H. and van der Vlugt L. S. (1934). *Water.*, 18: 147- 165.

Walsh, D., Pope, D., Danford, M. and Huff, T. (1993). The effect of microstructure on microbiologically influenced corrosion. *JOM.*, 45: 22- 30.

Wang, H. and Liang, C. H. (2007). Effect of Sulfate Reduced Bacterium on Corrosion Behaviour of 10CrMoAl Steel. *Journal of Iron and Steel Research, International.*, 14(1): 74- 78.

Wang, X., Melchers, R. E. and Loss J. (2017). *Prev. Process Ind.*, 45: 29- 42.

Warscheid, T. and Braams, J. (2000). Biodeterioration of stone: a review. *Int. Biodeterior. Biodegrad.*, 46: 343- 368.

Wu, T., Yan, M., Zeng, D., Xu, J., Sun, C., Yu, C., Ke, W. and Mater. J. (2015). *Sci. Technol.*, 31: 413- 422.

Xia, J., Yang, C., Xu, D., Sun, D., Nan, L., Sun, Z., Li, Q., Gu, T. and Yang, K. (2015). *Biofouling,* 31: 481- 492

Xu, D., Huang, W., Ruschau, G., Hornemann, J., Wen, J. and Gu, T. (2013). *Eng. Fail. Anal.*, 28: 149- 159.

Xu, D., Jia, R., Li, Y. and Gu, T. (2017). *World J. Microbiol. Biotechnol.*, 33 (2017) 97.

Xu, D., Li, Y., Song, F. and Gu, T. (2013). *Corros. Sci.*, 77: 385- 390.

Xu, D., Xia, J., Zhou, E., Zhang, D., Li, H., Yang, C., Li, Q., Lin, H., Li, X. and Yang, K. *Bioelectrochemistry.*, 113: 1- 8.

Zhang, P., Xu, D., Li, Y., Yang, K. and Gu, T. (2015). *Bioelectrochemistry.* 101: 14-21.

Zhao, K. (2009). *Investigation of microbiologically influenced corrosion (MIC) and biocide treatment in anaerobic salt water and development of a mechanistic MIC model.* Ohio University.

Zhou, E., Li, H., Yang, C., Wang, J., Xu, D., Zhang, D. and Gu, T. (2018). *Int. Biodeterior. Biodegrad.*, 127: 1- 9.

Zuo, R. (2007). Biofilms: strategies for metal corrosion inhibition employing microorganisms. *Microbiol. Biotechnol.*, 76: 1245- 1253.

Chapter 2

Abdallah, M., Benoliel, C., Drider, D., Dhulster, P. and Chihib, N. E. (2014). Biofilm formation and persistence on abiotic surfaces in the context of food and medical environments. *Arch. Microbiol.*, 196: 453- 472. doi: 10.1007/s00203- 014-0983- 1.

Beech, I. B. and Cheung, C. W. S. (1995). Interactions of exopolymers produced by sulphate-reducing bacteria with metal ions. *Int. Biodeterior. Biodegrad.*, 35: 59- 72.

Beech, W. B. and Sunner, J. (2004). Biocorrosion: towards understanding interactions between biofilms and metals. *Curr. Opin. Biotechnol.*, 15 (3): 181- 186.

Camargo, A. C., Woodward, J. J., Call, D. R. and Nero, L. A. (2017). *Listeria monocytogenes* in food-processing facilities, food contamination, and human listeriosis: the Brazilian Scenario. *Foodborne Pathog. Dis.*, 14: 623- 636. doi: 10.10 89/fpd.2016.2274.

Characklis, W. C. (1981). Fouling biofilm development: a process development. *Biotechnol. Bioeng.* 23: 1923- 1960. doi 10.1002/bit.260230902.

Characklis, W. C. and Marshall, K. C. (eds). (1990). *Biofilms.* John Wiley & Sons, New York., 796.

References

Claus, G. and Müller, R. (1996). Biofilms in a paper mill process water system. In *Microbially Influenced Corrosion of Materials,* (E. Heitz, H. C. Flemming and W. Sand, ed). Springer-Verlag, Berlin, Germany. 429- 437.

Colagiorgi, A., Bruini, I., Di Ciccio, P. A., Zanardi, E., Ghidini, S. and Ianieri, A. (2017). Listeria monocytogenes biofilms in the wonderland of food industry. *Pathogens.,* 6: E41. doi: 10.3390/pathogens6030041.

Costerton J. W., Lewandowski, Z., Caldwell, D. E., Korber, D. R., Lappin-Scott, H. M. (1995). Microbial biofilms. *Annu. Rev. Microbiol.,* 49711- 745.

Costerton, J. W. and Lappin-Scott, H. M. (1995). Introduction to microbial biofilms. In *Microbial Biofilms* (HM Lappin-Scott and JW Costerton, eds), Plant and microbial biotechnology research series: 5. University Press, Cambridge, UK., 1- 11.

Costerton, J. W., Cheng, K. J., Geesey, G. G., Ladd, T. I., Nickel, J. C., Dasgupta, M. and Marrie, T. J. (1987). Bacterial biofilms in nature and disease. *Annu. Rev. Microbiol.,* 41: 435- 464.

Coughlan, L. M., Cotter, P. D., Hill, C. and Álvarez-Ordóñez, A. (2016). New weapons to fight old enemies: novel strategies for the (bio)control of bacterial biofilms in the food industry. *Front. Microbiol.* 7: 1641. doi: 10.3389/fmicb.2016. 01641.

Davey, M. E. and O'Toole, G. A. (2000). Microbial biofilms: from ecology to molecular genetics. *Microbiology and Molecular Biology Reviews,* 64: 847- 867.

Davies, D. G., Chakrabarty, A. M. and Geesey, G. G. (1993). Exopolysaccharide production in biofilms: substratum activation of alginate gene expression by *Pseudomonas aeruginosa*. *Appl. Environ. Microbiol.,* 59(4): 1181- 1186.

Dunne, W. M. (2002). Bacterial adhesion: seen any good biofilms lately? *Clin. Microbiol. Rev.,* 15: 155- 166.

Edwards, M. and Sprague, N. (2001). Organic matter and copper corrosion by-product release: a mechanistic study. *Corros. Sci.,* 43(1): 1- 18.

Flemming, H. C. (1996). Economical and technical overview, In *Microbially influenced corrosion of materials,* (E. Heitz, H. C. Flemming and W. Sand, ed). Springer-Verlag, Berlin, Germany., 6- 14.

Flemming, H. C. and Wingender, J. (2010). The biofilm matrix. *Nat. Rev. Microbiol.,* 8: 623- 633.

Fox, E. P., Singh- babak, S. D., Hartooni, N. and Nobile, C. J. (2015). Biofilms and antifungal resistance. In: Coste, A. T., Vandeputte, P., editors. *Antifungals from Genomics to Resistance and the Development of Novel Agents.* Caister Academic Press., 71-90.

Galié, S., García-Gutiérrez, C., Miguélez, E. M., Villar, C. J. and Lombó, F. (2018). Biofilms in the Food Industry: Health Aspects and Control Methods. *Front. Microbiol.,* 9: 898. doi: 10.3389/fmicb.2018.00898.

Garrett, E. S., Perlegas, D. and Wozniak, D. J. (1999). Negative control of flagellum synthesis in *Pseudomonas aeruginosa* is modulated by the alternative sigma factor AlgT (AlgU). *J. Bacteriol.,* 181(23): 7401- 7404.

Heilmann, C., Hussain, M., Peters, G. and Götz, F. (1997). Evidence for autolysin-mediated primary attachment of *Staphylococcus epidermidis* to a polystyrene surface. *Mol. Microbiol.,* 24(5): 1013- 1024.

References

Huttunen-Saarivirta, E., Rajala, P., Marja-aho, M., Maukonen, J., Sohlberg, E. and Carpen, L. (2018). Ennoblement, corrosion, and biofouling in brackish seawater: Comparison between six stainless steel grades. *Bioelectrochemistry.* 120: 27-42.

Karatan, E. and Watnick, P. (2009). Signals, regulatory networks and materials that build and break bacterial biofilms. *Microbiol. Mol. Biol. Rev.,* 73: 310- 347.

Karn, S. K., Bhambri, A., Jenkinson, I. R., Duan, J. and Kumar, A. (2020). The roles of biomolecules in corrosion induction and inhibition of corrosion: a possible insight. *Corros. Rev.,* 1- 19. https://doi.org/10.1515/corrrev-2019-0111.

Karn, S. K., Chakrabarti, S. K. and Reddy, M. S. (2011). Degradation of pentachlorophenol by *Kocuria* sp. CL2 isolated from secondary sludge of pulp and paper mill. *Biodegardation.,* 22(1): 63- 69. ISSN: 0923-9820.

Korber, D. R., Lawrence, J. R., Lappin- Scott, H. M. and Costerton, J. W. (1995). Growth of microorganisms on surfaces. In *Microbial Biofilms* (HM Lappin-Scott and JW Costerton, ed), Plant and microbial biotechnology research series: 5. University Press, Cambridge, UK., 15- 45.

Koskinen, R., Ali-Vehmas, T., Kämpfer, P., Laurikkala, M., Tsitko, I., Kostyal, E., Atroshi, F. and Salkinoja-Salonen, M. S. (2000). Characterization of *Sphingomonas* isolates from the Finnish and Swedish drinking water distribution system. *J Appl Microbiol.,* 89: 687- 696.

LeChevallier, M. W. (1991). Biocides and the current status of biofouling control in water systems. In: *Biofouling and Biocorrosion in Industrial Water Systems* (H. C. Flemming and GG Geesey, ed). Springer-Verlag, Berlin, Germany., 114- 132.

LeChevallier, M. W. (1999). Biofilms in drinking water distribution systems: significance and control. In: *Identifying Future Drinking Water Contaminants.* National Research Council. National Academy Press, Washington, D.C., 206- 219.

Linhard, P. (1996). Failure of chromium nickel steel in a hydroelectric power plant by manganese-oxidizing bacteria. In *Microbially Influenced Corrosion of Materials,* (E Heitz, H-C Flemming and W Sand, ed). Springer-Verlag, Berlin, Germany., 221- 230.

Marshall, K. C. (1997). Colonization, adhesion, and biofilms. In: *Manual of Environmental Microbiology,* (CJ Hurst, GR Knudsen, MJ McInerney, LD Stetzenbach and MV Walter, eds). ASM Press, Washington D.C., USA., 358- 365.

Martínez, A., Torello, S. and Kolter, R. (1999). Sliding motility in mycobacteria. *J. Bacteriol.,* 181: 7331- 7338.

McBride, M. J. (2001). Bacterial gliding motility: multiple mechanisms for cell movement over surface. *Annu Rev Microbiol.,* 55: 49-75.

Meyer, R. L. (2015). Intra- and inter-species interactions within biofilms of important foodborne bacterial pathogens. *Front. Microbiol.,* 6: 841. doi:10.3389/fmicb.2015. 00841.

Mizan, M. F., Jahid, I. K. and Ha, S. D. (2015). Microbial biofilms in seafood: a food-hygiene challenge. *Food Microbiol.,* 49: 41- 55. doi: 10.1016/j.fm.2015. 01.009.

Mollica, A. (1992). Biofilm and corrosion on active-passive alloys in sea water. *Int. Biodeterior. Biodegrad.,* 29: 213–229.

Nielsen, A. T., Tolker-Nielsen, T., Barken, K. B. and Molin, S. (2000). Role of commensal relationships on the spatial structure of a surface-attached microbial consortium. *Environ Microbiol.*, 2: 59- 68.

Nikolaev, Y. A. and Plakunov, V. K. (2007). Biofilm - "City of Microbes" or an analogue of multicellular organisms? *Int. J. Mol. Sci.*, 76, 125- 138. doi: 10.1134/S00262617 07020014.

O'Toole, G. A. and Kolter, R. (1998). Flagellar and twitching motility are necessary for *Pseudomonas aeruginosa* biofilm development. *Mol. Microbiol.*, 30(2): 295- 304.

Otto, M. (2008). *Staphylococcal* biofilms. *Curr. Top. Microbiol. Immunol.*, 207–228.

Palmer, J., Flint, S. and Brooks, J. (2007). Bacterial cell attachment, the beginning of a biofilm. *J. Ind. Microbiol. Biotechnol.*, 34(9): 577- 588.

Pirttijärvi, T. S. M., Ahonen, L. M., Maunuksela, L. M. and Salkinoja-Salonen, M. S. (1998). *Bacillus cereus* in a whey process. *Int J Food Microbiol.*, 44: 31- 41.

Renner, L. D. and Weibel, D. B. (2011). Physicochemical regulation of biofilm formation. *MRS Bull.*, 36(5): 347–355.

Saxena, P., Joshi, Y., Rawat, K. and Bisht, R. (2018). Biofilms: Architecture, Resistance, Quorum Sensing and Control Mechanism. *Indian. J. Microbiol.*, 1- 10. https://doi.org/10.1007/s12088-018-0757-6.

Srey, S., Jahid, I. K. and Ha, S. (2013). Bio film formation in food industries: a food safety concern. *Food Control*. 31: 572- 585. doi: 10.1016/j.foodcont.2012.12.001.

Storgårds, E. (2000). Process hygiene control in beer production and dispensing. VTT publications 410, Ph.D. thesis. *University of Helsinki, Helsinki, Finland.*

Sutherland, I. (2001). Biofilm exopolysaccharides: a strong and sticky framework. *Microbiology.*, 147(1): 3- 9.

Szewzyk, U., Szewzyk, R., Manz, W. and Schleifer, K. H. (2000). Microbiological safety of drinking water. *Annu Rev Microbiol.*, 54: 81- 127.

Usher, K. M., Kaksonen, A. H., Cole, I. and Marney, D. (2014). Critical review: Microbially influenced corrosion of buried carbon steel pipes. *Int. Biodeterior. Biodegrad.*, 93: 84-106. doi: 10.1016/j.ibiod.2014.05.007.

Väisänen, O. M, Weber, A., Bennasar, A., Rainey, F. A., Busse, H. J. and Salkinoja-Salonen, M. S. (1998). Microbial communities of printing paper machine. *J ApplMicrobiol.*, 84: 1096- 1084.

Väisänen, O. M., Nurmiaho-Lassila, E. L., Marmo, S. A. and Salkinoja-Salonen, M. S. (1994). Structure and composition of biological slime on paper and board machines. *Appl. Environ. Microbiol.*, 60: 641- 653.

Vishwakarma, V. (2019). Impact of environmental biofilms: Industrial components and its remediation. *Journal of Basic Microbiology*. https://doi.org/10.1002/jobm.201900 569.

Wagner, D., Fischer, W. R., Chamberlain, A. H. L., Wardell, J. N. and Sequeira, C. A. C. (2004). Microbiologically Influenced Corrosion of coper in potable water installations-a European project review. *Materials and Corrosion.*, 48(5): 311-321.

Whittaker, C. J., Klier, C. M. and Kolenbrander, P. E. (1996). Mechanisms of adhesion by oral bacteria. *Annu Rev Microbiol.* 50: 513- 552.

Xie, H., Cook, G. S., Costerton, J. W., Bruce, G., Rose, T. M. and Lamont, R. J. (2000). Intergenic communication in dental plaque biofilms. *J Bacteriol.*, 182: 7067- 7069.

Chapter 3

Bharathi, P. A. L. (2008). Sulfur cycle. *Encyclopedia of Ecology* (Second Edition). 4: 192- 199. https://doi.org/10.1016/B978-0-444-63768-0.00761-7.

Dong, X., Kleiner, M., Sharp, C. E., Thorson, E., Li, C., Liu, D., & Strous, M. (2017). Fast and simple analysis of MiSeq amplicon sequencing data with Meta. *Amp. Front. Chem.* 8: 1461.

Dunlap, P. V. (2001). In E*ncyclopedia of Biodiv*ersity (Second Edition).

Ehrlich, H. L. and Newman, D. K. (2009). *Geomicrobiology.* CRC Press Taylor & Francis Group: Boca Raton, FL., 606.

Fakruddin, M. and Mannan, K. (2013). Methods for analyzing diversity of microbial communities in Natural Environments. *Ceylon Journal of Cience,* (Bio Sci). 42(1): 19- 33.

Fredrickson, J. K., Romine, M. F., Beliaev, A. S., Auchtung, J. M., Driscoll, M. E., Gardner, T. S., Nealson, K. H., Osterman, A. L., Pinchuk, G., Reed, J. L., Rodionov, D. A., Rodrigues, J. L. M., Saffarini, D. A., Serres, M. H., Spormann, A. M., Zhulin, I. B., & Tiedje, J. M. (2008). Towards environmental systems biology of *Shewanella. Nat Rev Microbiol.,* 6: 592- 603.

Gong, Y., Werth, C. J., He, Y., Su, Y., Zhang, Y. and Zhou, X. (2018). Intracellular versus extracellular accumulation of hexavalent chromium reduction products by *Geobactersulfurreducens* PCA. *Environ. Pollut.* 240: 485- 492.

Iverson, W. P. (1987). Microbial Corrosion of Metals. *Advances in Applied Microbiology.* 32: 1- 36.

Jurgens, G., Glöckner, F., Amann, R., Saano, A., Montonen, L., Likolammi, M., Münster U. (2003). Identification of novel Archaea in bacterioplankton of a boreal forest lake by phylogenetic analysis and fluorescent *in situ* hybridization. *FEMS Microbiol Ecol.,* 34(1): 45- 56.

Karn, S. K., Bhambri, A., Jenkinson, I. R., Duan, J. and Kumar, A. (2020). The roles of biomolecules in corrosion induction and inhibition of corrosion: a possible insight. *Corros. Rev.*, 1- 19. https://doi.org/10.1515/corrrev-2019-0111.

Karn, S. K., Chakrabarti, S. K. and Reddy, M. S. (2010a). Isolation and characterization of pentachlorophenol degrading *Bacillus* sp. isolated from secondary sludge of pulp and paper industry. *International Biodeterioration and Biodegradation.,* 64: 609- 63.

Karn, S. K., Chakrabarti, S. K. and Reddy, M. S. (2010b). Pentachlorophenol degradation by *Pseudomonas stutzeri* CL7 in the second art sludge of pulp and paper mill. *Journal of Environmental Sciences.,* 22(10): 1608- 1612.

Karn, S. K., Chakrabarti, S. K. and Reddy, M. S. (2011). Degradation of pentachlorophenol by *Kocuria* sp. CL2 isolated from secondary sludge of pulp and paper mill. *Biodegardation.,* 22 (1): 63- 69.

Karn, S. K., Fang, G. and Duan, J. (2017). *Bacillus* sp. act as dual role for corrosion induction and corrosion inhibition with carbon steel (CS). *Frontier in Microbiology.* 8: 1- 11.

Loreau, M. (2010). *From Populations to Ecosystems: Theoretical Foundations for a New Ecological Synthesis*. Princeton University Press, New Jersey, *USA.,* 320.

Luef, B., Fakra, S. C., Csencsits, R., Wrighton, K. C., Williams, K. H., Wilkins, M. J., Downing, K. H., Long, P. E., Comolli, L. R. and Banfield, J. F. (2013). Iron-reducing bacteria accumulate ferric oxyhydroxide nanoparticle aggregates that may support planktonic growth. *The ISME Journal.* 7: 338- 350.

Magurran, A. E. (2004). *Measuring Biological Diversity.* Blackwell: Oxford, UK.,264.

Mahadevan, R., Palsson, B. O. and Lovley, D. R. (2001). *In- situ* to in silico and back: elucidating the physiology and ecology of *Geobacter* sp. using genome-scale modelling. *Nat Rev.* 9: 39- 50.

Myers, C. R. and Kenneth, H. (1988). Nealson bacterial manganese reduction and growth with manganese oxide as the sole electron acceptor. *Science. New Series,* 240 (4857): 1319- 1321.

Nealson, K. H. (1997). Sediment bacteria: who's there, what are they doing, and what's new? *Annu Rev Earth Planetary Sci.*, 25: 403- 434.

Tebo, B. M., Bargar, J. R., Clement, B. G., Dick, G. J., Murray, K. J., Parker, D., Verity, R. and Webb, S. M. (2004). Biogenic manganese oxides: Properties and mechanisms of formation. *Annu Rev Earth Pl Sc.* 32: 287- 328.

Weber, K. A., Achenbach, L. A. and Coates, J. D. (2006). Microorganisms pumping iron: anaerobic microbial iron oxide. *Nature Reviews Microbiology*, 4: 752-764.

Chapter 4

Agrawal, P. K. and Shrivastava, R. (2013). Molecular Markers. In: Saxena, J., Ravi, I. and Bhaunthiyal, M. (eds) *Advance in Biotechnology.* Springer, 25- 39.

Aitken, M. D., Stringfellow, W. T., Nagel, R. D., Kazunga, C. and Chen, S. H. (1998). Characteristics of phenanthrene-degrading bacteria isolated from soils contaminated with polycyclic aromatic hydrocarbons. *Can. J. Microbiol.* 44: 743- 752.

Barac, T., Taghavi, S., Borremans, B., Provoost, A., Oeyen, L., Colpaert, J. V., Vangronsveld, J. and van der Lelie, D. (2004). Engineered endophytic bacteria improve phytoremediation of water-soluble, volatile, organic pollutants. *Nat. Biotechnol.* 22: 583-588.

Bodelier, P. L. E., Roslev, P., Henckel, T. and Fenzel, P. (2000). Stimulation by ammonium-based fertilizers of methane oxidation in soil around rice root. *Nature.* 40: 421- 424.

Brenner, S., Johnson, M., Bridgham, J., Golda, G., Lloyd, D. H., Johnson, D., Luo, S., McCurdy, S., Foy, M., Ewan, M., Roth, R., George, D., Eletr, S., Albrecht, G., Vermaas, E., Williams, S. R., Moon, K., Burcham, T., Pallas, M., Corcoran, K. (2000). Gene expression analysis by massively parallel signature sequencing (MPSS) on microbead arrays. *Nat. Biotechnol.*, 18: 630- 634.

Bruns, M. A., Stephen, J. R., Kowalchuk GA, Prosser, J. I. and Paul, E. A. (1999). Comparative diversity of ammonia oxidizer 16S rRNA gene sequences in native, tilled and successional soils. *Appl. Environ. Microbiol.* 65: 2994- 3000.

Bumpus, J. A. (1989). Biodegradation of polycyclic aromatic hydrocarbons by *Phanerochaetechrysosporium. Appl. Environ. Microbiol.*, 61: 2631- 2635.

References

Cahyani, V. R., Matsuya, K., Asakawa, S. and Kiumura, M. (2003). Succession and phylogenetic profile of eukaryotic committees in the compositing process of rice straw estimated by PCR-DGE analysis. *Biol. Fertil. Soil.* 40: 334- 344.

Chayani, V. R., Watanbe, A., Matsuya, K., Asakawa, S. and Kimura, M. (2001). Succession of microbiota estimated by phospholipids fatty acids analysis and changes in organic constituents during the composting process of rice straw. *Soil Sci Plant Nutr.,* 48: 143- 735.

Cheem, S. A., Khan, M. I., Shen, C., Tang, X., Farooq, M., Chen, L., Zhang, L. and Chen, Y. (2010). Degradation of phenanthrene and pyrene in spiked soils by single and combined plants cultivation. *J. Hazard. Mat.,* 177: 384- 389.

Cho, J. C. and Tiedje, J. M. (2001). Bacterial species determining from DNA– DNA hybridization by using genome fragments and DNA microarrays. *Appl. Environ. Microbiol.* 67: 3677- 3682.

Choudhary, D. K. (2011). First preliminary report on isolation and characterization of novel *acinetobacter* sp. in casing soil used for cultivation of button mushroom. *Agaricus bisporus* (Lange) Imbach. *Int. J. Microbiol.* 790285: 1- 6.

Choudhary, D. K., Agrawal, P. K. and Johri, B. N. (2009). Characterization of functional activity in composted casing amendments used in cultivation of *Agaricus bisporus* (Lange) Imbach. *Indian J. Biotechnol.* 8: 97- 109.

Clegg, C. D., Ritz, K. and Griffiths, B. S. (2000). % G? C profiling and cross hybridization of microbial DNA reveals great variation in below ground community structure in UK upland grasslands. *Appl. Soil Ecol.,* 14: 125- 134.

DeSantis, T. Z., Brodie, E. L., Moberg, J. P., Zubieta, I. X., Piceno, Y. M. and Andersen, G. L. (2007). High-density universal 16S rRNA microarray analysis reveals broader diversity than typical clone library when sampling the environment. *Microb. Ecol.,* 53: 371-383.

Droffner, M. L. and Brinton, W. F. (1995). Survival of E. coli and salmonella populations in aerobic thermophilic composts as measured with DNA gene probes. *ZentralblHygUmweltmed.,* 197: 387- 397.

Dunlap, P. V. (2001). Microbial Diversity. *Encyclopedia of Biodiversity.,* 4: 280- 291.

Eiland, F., Klamner, M., Lind, A., Leth, M., and Bath, E. (2001). Influence of initial C:N ratio on chemical and microbial composition during long term composting of straw. *Microb. Ecol.,* 41: 272- 280.

Fakruddin, M., Shahnewaj, K. and Mannan, B. (2013). Methods for analyzing diversity of microbial communities in natural environments. *Ceylon J Sci.,* 42(1): 19- 33.

Felske, A., Wolterink, A., Van Lis, R. and Akkermanns, A. D. L. (1998). Phylogeny of the main bacterial 16S rRNA sequences in Drentse A grassland soils. *Appl. Environ Microbiol* 64: 879.

Fisher, M. M. and Triplett, E. W. (1999). Automated approach for ribosomal intergenic spacer analysis of microbial diversity and its application to fresh water bacterial communities. *Appl. Enviorn. Microbiol.,* 65: 4630- 4636.

Frisli, T., Haverkamp, T. H., Jakobsen, K. S., Stenseth, N. C. and Rudi, K. (2013). Estimation of meta- genome size and structure in an experimental soil microbiota from low coverage next- generation sequence data. *J. Appl. Microbiol.,* 114: 141–151.

References

Gałązka, A. and Gałązka, R. (2015). Phytoremediation of polycyclic aromatic hydrocarbons in soils artificially polluted using plant-associated-endophytic bacteria and Dactylis glomerata as the bioremediation plant. *Pol. J. Microbiol.* 64: 239- 250.

Gałązka, A., Król, M. and Perzyński, A. (2010). Bioremediation of crude oil derivatives in soils naturally and artificially polluted with the use of maize as the test plant. Part II. *Crop yeld. Acta Sci. Pol. Agricultura.*, 9: 25- 36.

Gałązka, A., Król, M. and Perzyński, A. (2010). Bioremediation of crude oil derivatives in soils naturally and artificially polluted with the use of maize as the test plant. Part I. PAHs degradation. *Acta Sci. Pol. Agricultura.* 9: 13- 24.

Gentry, R. W., McCarthy, J., Layton, A., McKay, L., Williams, D., Koirala, S. R. and Sayler, G. S. (2006). Escherichia coli loading at or near base flow in a mixed-use watershed. *J. Environ. Qual.* 35: 2244- 2249.

Giovannoni, S. J., Britschgi, T. B., Moyer, C. L. and Field, K. G. (1990). Genetic diversity in Saragasso Sea Bacterioplankton. *Nature.*, 345: 60- 62.

Gomez, E., Bisaro, V. and Conti, M. (2000). Potential C-source utilization pattern of bacterial communities as influenced by clearing and land use in a vertic soil of Argentina. *Appl Soil Ecol.*, 15: 273- 281.

Gorgé, O., Bennett, E. A., Massilani, D., Daligault, J., Pruvost, M., Geigl, E. M., Grange, T. (2016). Analysis of ancient DNA in microbial ecology. *Methods Mol. Biol.*, 1399: 289- 315.

Goris, J., Konstantinidis, K. T., Klappenbach, J. A., Coenye, T., Vandamme, P. and Tiedje, J. M. (2007). DNA-DNA hybridization values and their relationship to whole-genome sequence similarities. *Int. J. Syst. Evol. Microbiol.*, 57: 81- 91.

Green, E. A. and Voordouw, G. (2003). Analysis of environmental microbial communities by reverse sample genome probing. *J. Microbiol. Methods.*, 53: 211- 219.

Griffiths, B. S., Ritz, K., Ebblewhite, N. and Dobson, G. (1999). Soil microbial community structure : effects of substrate loading rates. *Soil Biol. Biochem.* 31: 145- 153.

Gurtner, C., Heyrman, J., Pinar, G., Lubitz, W., Swings, J. and Rolleke, S. (2000). Comparative analyses of the bacterial diversity on two different bio-deteriorated wall paintings by DGGE and 16S rDNA sequence analysis. *Int. Biodeterior. Biodegrad.*, 46: 229- 239.

Hansgate, A. M., Schloss, R. D., Hay, A. G. and Walker, L. P. (2004). Molecular characterization of fungal community dynamics in the initial stages of composting. *FEMS Microbiol. Ecol.*, 51: 209- 316.

Heyndrickx, M., Vauterin, L., Vandamme, P., Kersters, K. and De Vos, P. (1996). Applicability of combined amplified ribosomal DNA restriction analysis (ARDRA) pattern in bacterial phylogeny and taxonomy. *J. Microbiol. Methods.* 26: 247- 259.

Hodkinson, B. P. and Grice, E. A. (2015). Next-generation sequencing: a review of technologies and tools for wound microbiome research. *Adv. Wound Care.*, 4: 50- 58.

Illumina. (2016). *An Introduction to Next-Generation Sequencing Technology.* 2016.

Ishii, K., Fukui, M. and Takii, S. (2000). Microbial succession during a composting process as evaluated by denaturing gradient gel electrophoresis analysis. *J. Appl. Microbiol.*, 89: 768- 777.

Koschinsky, S., Peters, S., Schwieger, F. and Tebbe, C. C. (1999). Applied molecular techniques to monitor microbial communities in composting processes. In: Bell C (ed) Progress in microbial ecology. *Proceedings of the International Symposium on Microbial ecology*, 8.

Kowalchuk, G. A., Naovmenko, Z. S., Derikx, P. J. L., Felske, A., Stephen, J. R. and Arkhipchenko, I. A. (1999). Molecular analysis of Ammonia-oxidizing bacteria of the b subdivision of the class proteobacteria in compost and composted materials. *Appl. Environ. Microbiol.* 65: 396- 403.

Leglize, P., Alain, S., Jacques, B. and Corinne, L. (2008). Adsorption of phenanthrene on activated carbon increases mineralization rate by specific bacteria. *J. Hazard. Mat.* 151: 339- 347.

Liu, W. T., Marsh, T. L., Cheng, H. and Forney, L. J. (1997). Characterization of microbial diversity by determining terminal restriction fragment length polymorphisms of genes encoding 16S rRNA. *Appl. Environ. Microbiol.*, 63: 4516- 4522.

McCaig, A. E., Glover, L. A. and Prosser, J. I. (2001). Numerical analysis of grassland bacterial community structure under different land management regimens by using 16S ribosomal DNA sequence data and denaturing gradient gel electrophoresis banding patterns. *Appl. Environ. Microbiol.* 67: 4554- 4559.

Muyzer, G., Dewall, E. C. and Uitterlinden, A. G. (1993). Profiling of complex bacterial population by denaturing gradient gel electrophoresis analysis of polymerase chain reaction-amplified genes coding for 16S rRNA. *Appl. Environ. Microbiol.* 59: 695- 700.

Muyzer, G. and Smalla, K. (1998). Application of denaturing gradient gel electrophoresis (DGGE) and temperature gradient gel electrophoresis (TGGE) in microbial ecology. *Antonie Leeuwenhoek.*, 73.

Narasimhan, K., Basheer, C., Bajic, V. B. and Swarup, S. (2003). Enhancement of plant-microbe interactions using a rhizosphere metabolomics-driven approach and its application in the removal of polychlorinated biphenyls. *Plant. Physiol.* 132: 146- 153.

Nicolaisen, M. H. and Ramsing, N. B. (2002). Denaturing gradient gel electrophoresis (DGGE) approaches to study the diversity of ammonia-oxidizing bacteria. *J. Microbiol. Methods* 50: 189- 203.

Nikolausz, M., Marialigeti, K. and Kovacs, G. (2004). Comparison of RNA and DNA-based species diversity investigations in rhizoplane bacteriology with respect to chloroplast sequence exclusion. *J. Microbiol. Methods.* 56: 365- 373.

Olsen, G. J., Lane, D. J., Giovannoni, S. J., Pace, N. R. and Stahl, D. A. (1986). Microbial ecology and evolution: a ribosomal RNA approach. *Annu. Rev. Microbiol.* 40: 337- 365.

Orita, M., Iwahana, H., Kanazawa, H., Hayashi, K. and Sekiya, T. (1989). Detection of polymorphism of human DNA by gel electrophoresis as single strand conformation polymorphisms. *Proc Natl Acad Sci USA.*, 86: 2766- 2770.

Pace, N. R. (1996). New perspective on the natural microbial world: molecular microbial ecology. *ASM News.,* 62: 463- 470.

Pace, N. R. (1999). Microbial ecology and diversity. *ASM News.,* 65: 328- 333.

Peters, S., Koschinsky, S., Schwieger, F. and Tebbe, C. C. (2000). Succession of microbial communities during hot composting as detected by PCR-single-strand-conformation-polymorphism-based genetic profiles of small-subunit rRNA genes. *Appl. Environ. Microbiol.* 66: 930- 936.

Rademaker, J. L. W. and de Bruijn, F. J. (1997). Characterization and classification of microbes by REP-PCR genomic fingerprinting and computer assisted pattern analysis. In: Caetano-Anolle's, G., Gresshoff, P. M. (eds) *DNA Markers: Protocols, Applications and Overviews.* John Wiley, New York, 51- 171

Ranjard, L., Brothier, E. and Nazaret, S. (2000). Sequencing bands of ribosomal intergenic spacer analysis fingerprints for characterization and microscale distribution of soil bacterium populations responding to mercury spiking. *Appl. Environ. Microbiol.,* 66: 5334- 5339.

Rawat, S. and Johri, B. N. (2014). Thermophilic fungi: diversity and significance in composting. *Kavaka.,* 42: 52- 68.

Rawat, S., Agrawal, P. K., Choudhary, D. K. and Johri, B. N. (2005). *Microbial Diversity and Community Dynamics of Mushroom Compost Ecosystem.* In: Satyanarayana, T., Johri, B. N. (eds) I. K. International Pvt. Ltd. New Delhi, 1027.

Romero, M. C., Cazau, M. C., Giorgieri, S. and Arambarri, A. M. (1998). Phenanthrene degradation by microorganisms isolated from a contaminated stream. *Environ. Pollut.* 101: 355- 359.

Saison, C., Derange, V., Oliver, R., Millard, P., Commeaux, C., Montage, D. and Le, Roux X. (2005). Alteration and resilience of the soil microbial community following compost amendment: effects of compost level and compost born microbial community. *Environ Microbiol.* 8: 247- 257.

Samarajeewa, A. D., Hammad, A., Masson, L., Khan, I. U., Scroggins, R. and Beaudette, L. A. (2014). Comparative assessment of next-generation sequencing, denaturing gradient gel electrophoresis, clonal restriction fragment length polymorphism and cloning- sequencing as methods for characterizing commercial microbial consortia. *J. Microbiol. Methods* 108: 103- 111.

Schuster, S. C. (2008). Next-generation sequencing transforms today's biology. *Nat. Met.,* 5: 16–18.

Singh, S. K., Vijay, B. and Mediratta, V. (2005). Molecular characterization of Humicola grisea isolates associated with *Agaricus bisporus* compost. *Curr Sci.* 10: 1745- 1749.

Smit, E., Leeflang, P. and Wernars, K. (1997). Detection of shifts in microbial community structure and diversity in soil caused by copper contamination using amplified ribosomal DNA restriction analysis. *FEMS Microbiol. Ecol.,* 23: 249- 261.

Song, J., Weon, H. Y., Yoon, S. H., Park, D. S., Go, S. J. and Suh, J. W. (2001). Phylogenetic diversity of thermophilic *actinomycetes* and *Thermoactinomyces* sp. isolated from mushroom composts in Korea based on 16S rRNA gene sequence analysis. *FEMS Microbiol. Lett.* 202: 97- 102.

Spitaels, F., Wieme, A. D. and Vandamme, P. (2016). MALDI-TOF MS as a Novel Tool for Dereplication and Characterization of Microbiota in Bacterial Diversity Studies.

In book: *Applications of Mass Spectrometry in Microbiology*. doi: 10.1007/978-3-319-26070-9_9.

Swofford, D. L., Olsen, G. J., Waddell, P. J. and Hillis, D. M. (1996). Phylogenetic inference. In: Hillis, D. M., Moritz, C. and Mable, B. K. (eds) *Molecular systematics*. Sinauer, Sunderland., 407- 514.

Takaku, H., Kodaira, S., Kimoto, A., Nashimoto, M. and Takagi, M. (2006). Microbial communities in the garbage composting with rice hull as an amendment revealed by culture-dependent and-independent approaches. *J. Biosci. Bioeng.* 101(1): 42- 50.

Theron, J. and Cloete, T. E. (2000). Molecular techniques for determining microbial diversity and community structure in natural environments. *Crit. Rev. Microbiol.* 26: 37-57.

Tiquia, S. M., Ichida, J. M., Keener, H. M., Elwell, D. L., Beut, E. H., Jr. Michel, F. C. (2005). Bacterial community profiles on feathers during composting as determined by terminal restriction fragment length polymorphism analysis of 16S rDNA genes. *Environ. Biotechnol.* 67: 412- 419.

Torsvik, V., Overas, L. and Thingstad, T. (2002). Prokaryotic diversity magnitude, dynamics, and controlling factors. *Science.* 296: 1064- 1066.

White, T. J., Bruns, T., Lee, S. and Taylor, J. (1990). Amplification and direct sequencing of fungal ribosomal RNA genes for phylogenetics. In: Innis, M. A., Gelfand, D. H., Sninsky, J. J. and White, T. J. (eds). *PCR Protocols: a guide to methods and applications*. Academic Press, San Diego, 315- 322.

Williams, J. G. K., Kubelik, A. R., Livak, K. J., Rafalski, J. A. and Tingey, S. V. (1990). DNA polymorphisms amplified by arbitrary primers are useful as genetic markers. *Nucleic Acids Res.*, 18: 6531- 6535.

Woese, C. R. (1987). Bacterial evolution. *Microbiol Rev.*, 51: 221- 271.

Yrjälä, K., Keskinen, A. K., Lkerman, M. L., Fortelius, C. and Sipila T. P. (2010). The rhizosphere and PAH amendment mediate impacts on functional and structural bacterial diversity in sandy peat soil. *Environ. Pollut.* 158: 1680- 1688.

Zoetendal, E. G., Cheng, B., Koike, S. and Mackie, R. I. (2004). Molecular microbial ecology of the gastrointestinal tract: from phylogeny to function. *Curr. Issues Intest. Microbiol.* 5: 31- 48.

Chapter 5

Ahmad, N. H., Mustafa, S. and Che Man, Y. B. (2015). Microbial polysaccharides and their modification approaches: a review. *Int. J. Food.Prop.*18(2): 332- 347.

Alves, V. D., Freitas, F., Torres, C. A. V., Cruz, M., Marques, R., Grandfils, C., Gonçalves, M. P., Oliveira, R., & Reis, M. A. M. (2009). Rheological and morphological characterization of the culture broth during exopolysaccharide production by *Enterobacter* sp. *Carbohydr. Polym.* doi: 10.1016/j.carbpol.2009.09.006.

Antón, J., Meseguer, I. and Rodriguez-Valera, F. (1988). Production of an extracellular polysaccharide by *Haloferaxmediterranei*. *Appl. Environ. Microbiol.*, 54(10): 2381- 2386.

References

Bajaj, I. B., Survase, S. A., Saudagar, P. S. and Singhal, R. S. (2007). Gellan gum: fermentative production, downstream processing and applications. *Food Technol. Biotech.*, 45(4): 341.

Bautista, B. E. T., Wikieł, A. J., Datsenko, I., Vera, M., Sand, W., Seyeux, A. Z. S., Frateur, I., and Marcus, P. (2015). Influence of extracellular polymeric substances (EPS) from *Pseudomonas* NCIMB 2021 on the corrosion behavior of 70 Cu–30 Ni alloy in seawater. *J. Electroanal. Chem.*, 737: 184- 197.

Beech, I. and Sunner, J. (2004). Biocorrosion: Towards understanding interactions between biofilms and metals. *Curr. Opin. Biotechnol.*, 15: 181- 186.

Beech, I. B. (2004). Corrosion of technical materials in the presence of biofilms - current understanding and state-of-the art methods of study. *Int. Biodeterior. Biodegrad.* 53: 177- 183.

Beech, I. B. and Coutinho, C. L. M. (2003). Biofilms on corroding materials. In: Lens, Moran, P., Mahony, A. P., Stoodly, T. and O'Flaherty, V. Eds. Biofilms in medicine, industry and environmental biotechnology-characteristics, analysis and control. *IWA Publication Alliance* House, London. 115- 131.

Beech, I. B. and Sunner, J. (2004). Biocorrosion: towards understanding interactions between biofilms and metals. *Curr. Opin. Biotechnol.*, 15: 181- 186.

Beech, I. B., Zinkevich, V., Tapper, R. and Gubner, R. (1998). Direct involvement of an extracellular complex produced by a marine sulfate reducing bacterium in deterioration of steel. *Geomicrobiol. J.*, 15: 121- 134.

Boonchai, R., Kaewsuk, J. and Seo, G. (2014). Effect of nutrient starvation on nutrient uptake and extracellular polymeric substance for microalgae cultivation and separation. *Desalination Water Treat.*, 55: 360- 367. doi: 10.1080/19443994.2014.939501

Branda, S. S., Vik, S., Friedman, L. and Kolter, R. (2005). Biofilms: The matrix revisited. *Trends. Microbiol.*, 13: 20- 26.

Busalmen, J. P., Vazquez, M. and De Sanchez, S. R. (2002). New evidences on the catalase mechanism of microbial corrosion. *Electrochim. Acta.*, 47: 1857- 1865.

Conlisk, A. T. (2013). Essentials of micro- and nano-fluidics. *Cambridge University Press.*, Cambridge, UK.

Dogsa, I., Kriechbaum, M., Stopar, D. and Laggner, P. (2005). Structure of bacterial extracellular polymeric substances at different pH values as determined by SAXS. *Biophysical Journal.* 89(4): 2711- 2720.

Dong, Z. H., Liu, T. and Liu, H. F. (2011). Influence of EPS isolated from thermophilic sulphate-reducing bacteria on carbon steel corrosion. *Biofouling.*, 27: 487- 495.

Elisashvili, V. I., Kachlishvili, E. T. and Wasser, S. P. (2009). Carbon and nitrogen source effects on basidiomycetes exopolysaccharide production. *Appl. Biochem. Microbiol.*, 45: 531- 535. doi: 10.1134/s0003683809050135

Feio, M. J., Zinkevich, V., Beech, I. B., Llobet-Brossa, E., Eaton, P., Schmitt, J. and Guezennec, J. (2004). *Desulfovibrioalaskensis* sp. nov., a sulphate-reducing bacterium from a soured oil reservoir. *Int. J. Syst. Evol. Microbiol.*, 54: 1747- 1752.

Fleming, K. G. (2000). Riding the wave: structural and energetic principles of helical membrane proteins. *Curr. Opin. Chem. Biol.* 11(1): 67- 71.

References

Flemming, H. C. and Wingender, J. (2010). The biofilm matrix. *Nat. Rev. Microbiol.*, 8: 623- 633. doi: 10.1038/nrmicro2415.

Freitas, F., Alves, V. D. and Reis, M. A. (2011). Advances in bacterial exopolysaccharides: from production to biotechnological applications. *Trends Biotechnol. Res.*, 29(8): 388- 398. doi: 10.1016/j.tibtech.2011.03.008.

Görke, B. and Stülke, J. (2008). Carbon catabolite repression in bacteria: many ways to make the most out of nutrients. *Nat Rev Microbiol.*, 6(8): 613- 624. doi: 10.1038/nrmicro1932.

Gutierrez, T., Berry, D., Yang, T., Mishamandani, S., McKay, L., Teske, A. and Aitken, M. D. (2013). Role of Bacterial Exopolysaccharides (EPS) in the Fate of the Oil Released during the Deepwater Horizon Oil Spill. *PLoS ONE.* 8: e67717.

Gutierrez, T., Morris, G. and Green, D. H. (2009). Yield and physicochemical properties of EPS from *Halomonas* sp. strain TG39 identifies a role for protein and anionic residues (sulfate and phosphate) in emulsification of n-hexadecane. *Biotechnol. Bioeng.* 103: 207- 216.

Gutierrez, T., Mulloy, B., Black, K. and Green, D. H. (2007). Glycoprotein emulsifiers from two marine *Halomonas* species: Chemical and physical characterization. *J. Appl. Microbiol.*,103: 1716- 1727.

Hamilton, W. A. (1985). Sulphate-reducing bacteria and anaerobic corrosion. *Annu. Rev. Microbiol.*, 39: 195- 217.

Hauser, L. J., Land, M. L., Brown, S. D., Larimer, F., Keller, K. L., Rapp-Giles, B. J., Price, M. N., Lin, M., Bruce, D. C. and Detter, J. C. (2011). Complete genome sequence and updated annotation of *Desulfovibrioalaskensis* G20. *J. Bacteriol.*, 193: 4268- 4269.

Hwang, H. J., Kim, S. W., Xu, C. P., Choi, J. W. and Yun, J. W. (2004). Morphological and rheological properties of the three different species of basidiomycetes Phellinus in submerged cultures. *J. Appl. Microbiol.*, 96: 1296- 1305. doi: 10.1111/j.1365-2672.2004.02271.x

Iijima, M., Shinozaki, M., Hatakeyama, T., Takahashi, M. and Hatakeyama, H. (2007). AFM studies on gelation mechanism of xanthan gum hydrogels. *Carbohydr. Polymers.* 68(4): 701- 707.

Jain, R., Raghukumar, S., Tharanathan, R. and Bhosle, N. B. (2005). Extracellular polysaccharide production by Thraustochytrid protists. *Mar. Biotechnol.*, 7: 184- 192. doi: 10.1007/s10126-004-4025-x.

Jenkinson, I. R. (2014). *Copepods: diversity, habitat and behavior.* Seuront, L. (Ed.). Nova Science Publishers, Inc, New York, 181- 214.

Kambourova, M., Mandeva, R., Dimova, D., Poli, A., Nicolaus, B. and Tommonaro, G. (2009). Production and characterization of a microbial glucan, synthesized by *Geobacillustepidamans*V264 isolated from Bulgarian hot spring. *Carbohydr. Polymers.* 77(2): 338- 343.

Karn, S. K., Fang, G. and Duan, J. (2017). *Bacillus* sp. act as dual role for corrosion induction and corrosion inhibition with carbon steel (CS). *Frontier in Microbiology.* 8: 1- 11. ISSN: 1664-302X. doi: 10.3389/fmicb.2017.02038.

Khani, M., Bahrami, A. and Ghafari, M. D. (2015). Optimization of operating parameters for anti-corrosive biopolymer production by *ChryseobacteriumIndologenes* MUT 2 using central composite design methodology. *J. Taiwan. Inst. Chem. Eng.* doi: 10.1016/j.jtice.2015.09.016

Kilbane, J., Bogan, B. and Lamb, B. (2005). Quantifying the contribution of various bacterial groups to microbiologically influenced corrosion. In *Proceedings of the NACE International Corrosion Conference,* Houston, TX, USA., 3- 7.

Kim, S. W., Xu, C. P., Hwang, H. J., Choi, J. W., Kim, C. W. and Yun, J. W. (2003). Production and Characterization of Exopolysaccharides from an Enthomopathogenic Fungus *Cordycepsmilitaris*NG3. *Biotechnol Prog.*, 19(2): 428- 435. doi: 10.1021/bp 025644k.

Kinzler, K., Gehrke, T., Telegdi, J. and Sand, W. (2003). Bioleaching - a result of interfacial processes caused by extracellular polymeric substances (EPS). *Hydrometallurgy.,* 71: 83- 88.

Kolter, R. and Greenberg, E. P. (2006). The superficial life of microbes. *Nature.* 441: 300- 302.

Kumar, S. A., Mody, K. and Jha, B. (2007). Bacterial exopolysaccharides–a perception. *J. Basic Microbiol.,* 47(2): 103- 117. doi: 10.1002/jobm.200610203.

Lapaglia, C. and Hartzell, P. L. (1997). Stress-Induced Production of Biofilm in the hyperthermophile *Archaeoglobusfulgidus. Appl. Environ. Microbial.,* 63(8): 3158- 3163.

Lee Chang, K. J., Nichols, C. M., Blackburn, S. I., Dunstan, G. A., Koutoulis, A., and Nichols, P. D. (2014). Comparison of thraustochytrids *Aurantiochytrium* sp., *Schizochytrium* sp., *Thraustochytrium* sp., and *Ulkenia* sp. for production of biodiesel, long-chain omega-3 oils, and exopolysaccharide. *Mar. Biotechnol.,* 16: 396–411. doi: 10.1007/s10126-014-9560-5

Li, W. W. and Yu, H. Q. (2014). Insight into the roles of microbial extracellular polymer substances in metal biosorption. *Bioresour Technol.*, 160: 15- 23. doi: 10.1016/j.bior tech.2013.11.074.

Lopez, J. C., Pérez, J. S., Sevilla, J. F., Fernandez, F. A., Grima, E. M. and Chisti, Y. (2003). Production of lovastatin by *Aspergillus terreus*: effects of the C: N ratio and the principal nutrients on growth and metabolite production. *Enzyme Microb Technol.,* 33(2): 270- 277. doi: 10.1016/S0141-0229(03)00130-3.

Morin, A. (1998). Screening of polysaccharide-producing microorganisms, factors influencing the production and recovery of microbial polysaccharides. In *Polysaccharides: Structural Diversity and Functional Versatility.* New York: Marcel Dekker Inc., 275- 296.

Morris, V. J., Gromer, A., Kirby, A. R., Bongaerts, R. J. and Gunning, A. P. (2011). Using AFM and force spectroscopy to determine pectin structure and (bio) functionality. *Food Hydrocoll.,* 25(2): 230- 237.

Nicolaus, B., Lama, L., Esposito, E., Manca, M. C., Improta, R., Bellitti, M. R. and Gambacorta, A. (1999). *Haloarcula* sp. able to biosynthesize exo- and endopolymers. *J. Ind. Microbiol Biotechnol.,* 23(6): 489- 496.

Nicolaus, B., Manca, M. C., Romano, I., and Lama, L. (1993). Production of an exopolysaccharide from two thermophilic archaea belonging to the genus Sulfolobus. *FEMS. Microbiol. Lett.*, 109(2-3): 203- 206.

Nwodo, U., Green, E. and Okoh, A. I. (2012). Bacterial exopolysaccharides: functionality and prospects. *Int. J. Mol. Sci.*, 13(11): 14002- 14015.

Paramonov, N., Parolis, H., Onnton, and Rodrguez-Valera, F. (1998). *The structure of the exocellular polysaccharide produced by the Archaeon Haloferaxgibbonsii ATCC 33959*. Carbohydr. Res., 309(1): 89- 94.

Parikh, A. and Madamwar, D. (2006). Partial characterization of extracellular polysaccharides from cyanobacteria. *Bioresour. Technol.*, 97: 1822-1827. doi: 10.10 16/j.biortech.2005.09.008.

Pavlova, K. and Grigorova, D. (1999). Production and properties of exopolysaccharide by *Rhodotorulaacheniorum* MC. *Food Res. Int.*, 32: 473- 477. doi: 10.1016/s0963-9969(99)00110-6.

Rinker, K. D. and Kelly, R. M. (1996). Growth physiology of the hyperthermophilic archaeon Thermococcus litoralis: Development of a sulfur-free defined medium, characterization of an exopolysaccharide and evidence of biofilm formation. *Appl. Environ. Microbiol.* 62(12): 4478- 4485.

Roca, C., Alves, V. D., Freitas, F., and Reis, M. A. (2015). Exopolysaccharides enriched in rare sugars: bacterial sources, production, and applications. *Front. Microbiol.*, 6: 288. doi: 10.3389/fmicb.2015.00288.

Rohwerder, T., Gehrke, T., Kinzler, K. and Sand, W. (2003). Bioleaching review part A: progress in bioleaching: fundamentals and mechanisms of bacterial metal sulfide oxidation. *Appl. Microbiol. Biotechnol.*, 63: 239- 248.

Sand, W. (2003). Microbial life in geothermal waters. *Geothermics*. 32: 655- 667.

Simsek, S., Mert, B., Campanella, O. H. and Reuhs, B. (2009). Chemical and rheological properties of bacterial succinoglycan with distinct structural characteristics. *Carbohydr. Polymers.*, 76(2): 320- 324.

Sutherland, I. W. (1985). Biosynthesis and Composition of Gram-Negative Bacterial Extracellular and Wall Polysaccharides. *Annu. Rev. Microbiol.*, 39: 243- 270.

Sutherland, I. W. (2001). The biofilm matrix-an immobilized but dynamic microbial environment. *Trends Microbiol.*, 9: 222- 227.

Vardharajula, S. and Ali, S. Z. (2015). The production of exopolysaccharide by *Pseudomonas putida* GAP-P45 under various abiotic stress conditions and its role in soil aggregation. *Microbiology.*, 84: 512- 519. doi: 10.1134/s0026261715040153.

Wang, X., Sharp, C. E., Jones, G. M., Grasby, S. E., Brady, A. L., & Dunfield, P. F. (2015). Stable-isotope probing identifies uncultured planctomycetes as primary degraders of a complex heteropolysaccharide in soil. *Appl. Environ. Microbiol.*, 81: 4607- 4615. doi: 10.1128/aem.00055-15.

Wikieł, A. J. (2013). *Role of extracellular polymeric substances on biocorrosion initiation or inhibition*. PhD. Universität Duisburg- Essen, Germany.

Wingender, J., Neu, T. R. and Flemming, H. C. (1999). What are Bacterial Extracellular Polymeric Substances? In *Microbial Extracellular Polymeric Substances*. Eds. Wingender, J. and Neu, T. R. Springer: Heidelberg, Germany. 1- 19.

Zinkevich, V., Bogdarina, I., Kang, H., Hill, M. A. W., Tapper, R. and Beech, I. B. (1996). Characterization of exopolymers produced by different isolates of marine sulphate-reducing bacteria. *Int. Biodeterior. Biodegrad.*, 37: 163- 172.

Chapter 6

Amaya, H. and Miyuki, H. (1995). Development of accelerated evaluation method for influenced corrosion resistance of stainless steels. *Corros. Eng.,* 44: 123- 133.

Amaya, H. and Miyuki, H. (1999). Laboratory reproduction of potential ennoblement of stainless steels in natural seawater. *Proceedings of Corrosion.* NACE: Houston, TX., 168.

Awad, M. I., Saad, A. F., Shaaban, M. R, Jahdaly, B. A. A. L. and Hazazi, O. A. (2017). New insight into the mechanism of the inhibition of corrosion of mild steel by some amino acids. *Int. J. Electrochem. Sci.,* 12: 1657- 1669.

Basseguy, R., Idrac, J., Jacques, C., Bergel, A., Delia, M. L. and Etcheverry, L. (2004). Local analysis by SVET of the involvement of biological systems in aerobic biocorrosion. In: Proceedings of Eurocorr. *International Scientific Committee for European Federation of Corrosion, Nice, France.*

Basséguy, R., Idrac, J., Jacques, C., Bergel, A., Delia, M. L. and Etcheverry L. (2004). Local analysis by SVET of the involvement of biological systems in aerobic biocorrosion. *Proceedings of Eurocorr, Nice, France.*

Beech, I. B. (2002). Biocorrosion: role of sulphate-reducing bacteria. In: Bitton, G. (Ed.). *Encyclopaedia environmental microbiology.,* 465- 475.

Beech, I. B. (2003). Sulfate-reducing bacteria in biofilms on metallic materials and corrosion. *Microbiol. Today.,* 30: 115- 117.

Beech, I. B. (2004). Corrosion of technical materials in the presence of biofilms - current understanding and state-of-the art methods of study. *Int. Biodeterior. Biodegrad.,* 53: 177- 183.

Beech, I. B. and Cheung, C. W. S. (1995). Interaction of exopolymers produced by sulfate-reducing bacteria with metal ions. *Int. Biodeterior. Biodegrad.* 35: 59- 72.

Beech, I. B. and Coutinho, C. L. M. (2003). Biofilms on corroding materials. In: Lens, P., Moran, A. P., Mahony, T., Stoodly, P. and O'Flaherty, V. (Eds.). *Biofilms in Medicine, Industry and Environmental Biotechnology-Characteristics, Analysis and Control.* IWA Publication Alliance House, London., 115- 131.

Blanco, A. and Blanco, G. (2017). Proteins. *Medical Biochemistry.,* 21- 71. https://doi.org/10.1016/B978-0-12-803550-4.00003-3.

Busalmen, J. P., Frontini, M. A. and de-Sanchez, S. R. (1998). Microbial corrosion: effect of the microbial catalase on the oxygen reduction. In: Walsh, F. C. and Campbell, S. A., (Ed.), Recent Karn, S. K., et al. *Biomolecules in Corrosion Induction and Inhibition Developments in Marine Corrosion.* Royal Society of Chemistry City, London, United Kingdom.

Busalmen, J. P., Vazquez, M. and De Sanchez, S. R. (2002). New evidences on the catalase mechanism of microbial corrosion. *Electrochim. Acta.,* 47: 1857- 1865.

Campbell, S., Gesey, G., Lewandowski, Z. and Jackson, G. (2004). Influence of the distribution of the manganese-oxidizing bacterium, *LeptothrixDiscophora*, on ennoblement of type 316 L stainless steel. *Corrosion.,* 60: 670.

Chandrasekaran, P. and Dexter, S. C. (1990). Bacterial metabolism in biofilm consortia: consequences for potential ennoblement. *Proceedings of Corrosion.* NACE: Houston, TX., 276.

Chandrasekaran, P. and Dexter, S. C. (1993). Mechanism of corrosion potential, ennoblement on passive alloys by seawater biofilms. In: *Proceedings of Corrosion*, NACE, International, Houston., 93: 493.

Chandrasekaran, P. and Dexter, S. C. (1994). In: *Proceedings of Corrosion*. NACE, Houston, TX, 276.

Chongdar, S., Gunasekaran, G. and Kumar, P. (2005). Corrosion inhibition of mild steel by aerobic biofilm. *Acta Electrochim.*, 50: 4655- 4665.

Costerton, J. W. and Boivin, J. (1991). In *Biofouling and Biocorrosion in Industrial Water Systems; Flemming*, H. C., Geesey, G. G. Eds. Springer: Berlin, Heilderberg.

Costerton, J. W., Lewandowski, Z., deBeer, D., Caldwell, D. E., Korber, D. R. (1994). James, G. Biofilms, the customised microniche. *J. Bacteriol.*, 1994: 176- 2137.

DaSilva, S., Basseguy, R. and Bergel, A. (2004). Electron transfer between hydrogenase and 316L stainless steel: identification of a hydrogenase-catalyzed cathodic reaction in anaerobic MIC. *J. Electroanal. Chem.*, 561: 93- 102.

Davidova, I. A., Duncan, K. E., Perez-Ibarra, B. M. and Suflita, J. M. (2012). Involvement of thermophilic archaea in the biocorrosion of oil pipe lines. *Environ. Microbiol.* 14: 1762- 1771.

De Bont, J. A. M. (1976). Hydrogenase activity in nitrogen fixing methane- oxidizing bacteria. *Anton Leeuw.* 42: 255- 259.

De Kairuz, M. S. N., Olazabal, M. E., Oliver, G., De Ruiz Holgado, A. A. P., Massa, E. and Farias, R. N. (1988). Fatty acid dependent hydrogen peroxide production in lactobacillus. *Biochem. Biophys. Res. Commun.*, 152: 113.

De la Rosa, C. and Yu, T. (2005). Three-dimensional mapping of oxygen distribution in wastewater biofilms using an automation system and microelectrodes. *Environ. Sci. Technol., 39*: 5196- 5202.

Denis, M., Arnaud, S. and Malatesta, F. (1989). Hydrogen peroxide is the end product of oxygen reduction by the terminal oxidase in the marine bacterium *Pseudomonas nautica* 617. *FEBS Lett.,* 247: 475.

De Silva Muñoz, L., Bergel, A. and Basseguy, R. (2007). Role of the reversible electrochemical deprotonation of phosphate species in anaerobic biocorrosion of steels. *Corrosion Sci.,* 49: 3998- 4004.

Deutzmann, J. S., Sahin, M. and Spormann, A. M. (2015). Extracellular enzymes facilitate electron uptake in biocorrosion and bio-electrosynthesis. *mBio.* 6: e00496-15.

Dexter, S. C. and Zhang, A. J. (1991). Effect of biofilms sunlight and salinity on corrosion potential and corrosion initiation of stainless alloy, EPRI NP-7275 final report on project 2939–4. *Electric Power Research Institute, Palo Alto, CA.*

Dickinson, W. H. and Lewandowski, Z. (1996). Manganese biofouling and the corrosion behaviour of stainless steel. *Biofouling.*, 10: 79- 93.

Dickinson, W. H., Lewandowski, Z. and Geer, R. D. (1996). Evidence for surface changes during ennoblement of type 316L stainless steel: dissolved oxidant and capacitance measurements. *Corrosion.*, 52.

Dinh, H. T., Kuever, J., Mussmann, M., Hassel, A. W., Stratmann, M. and Widdel, F. (2004). Iron corrosion by novel anaerobic microorganisms. *Nature.*, 427: 829- 832.

Dubiel, M., Hsu, C. H., Chien, C. C., Mansfeld, F. and Newman, D. K. (2002). Microbial iron respiration can protect steel from corrosion. *Appl. Environ. Microbiol.*, 68: 1440- 1445.

Dupont, I., Féron, D. and Novel, G. (1997). Influence des facteursinorganiques sur l'évolution du potentiel des aciersinoxydableseneau de mer naturelle [*Influence of organic factors on the evolution of the potential of stainless steels in natural seawater*]. *Matér. Techniq.*, 11: 41.

Dupont, I., Feron, D. and Novel, G. (1998). Effect of glucose oxidase ´ activity on corrosion potential of stainless steels in seawater. *Int. Biodeterior. Biodegrad.*, 41: 13-18.

Ece, G. G. and Bilgiç, S. (2010). A theoretical study on the inhibition efficiencies of some amino acids as corrosion inhibitors of nickel. *Corrosion Sci.*, 52: 3435- 3443.

Enzyme Nomenclature, IUBMB, Academic Press: San Diego, CA, 1992 (with Supplement 1 (1993), Supplement 2 (1994), Supplement 3 (1995), Supplement 4 (1997) and Supplement 5. In *Eur. J. Biochem.* 1994, 223, 1–5; *Eur. J. Biochem.* 1995, 232, 1–6; *Eur. J. Biochem.* 1996, 237, 1–5; *Eur. J. Biochem.* 1997, 250; 1–6, and *Eur. J. Biochem.* 1999, 264, 610–650; respectively.)

Franklin, M. J. and White, D. C. (1991). Biocorrosion. *Curr. Opin. Biotechnol.*, 2: 450.

Fu, J. J., Li, S. N., Cao, L. H., Wang, Y., Yan, L. H. and Lu, L. D. (2010). L-Tryptophan as green corrosion inhibitor for low carbon steel in hydrochloric acid solution. *J. Mater. Sci.*, 45: 979- 986.

Geiser, M., Avci, R. and Lewandowski, Z. (2002). Initiated pitting on 316L stainless steel. *Int. Biodeterior. Biodegrad.* 49: 235- 243.

Holthe, R., Bardal, E. and Garland, P. O. (1989). Time dependence of cathodic properties of materials in seawater. Stainless steel, titanium, platinum and 90/10 CuNi. *Mater. Perf.*, 28: 16.

Hutchinson, G. E. (1961). The paradox of the plankton. *Am. Nat.*, 95: 137- 145.

Ishii, S., Suzuki, S., Norden-Krichmar, T. M., Phan, T., Wanger, G., Nealson, K. H., Sekiguchi, Y., Gorby, Y. A. and Bretschger, O. (2014). Microbial population and functional dynamics associated with surface potential and carbon metabolism. *ISME J.*, 8: 963- 978.

Iverson, I. P. and Olson, G. J. (1983). Anaerobic corrosion by sulfate- reducing bacteria due to highly reactive volatile phosphorus compound. In: *Microbial corrosion.* Metals Society, London., 46.

Jayaraman, A., Ornek, D., Duarte, D. A., Lee, C. C., Mansfeld, F. B. and Wood, T. K. (1999). Axenic aerobic biofilms inhibit corrosion of copper and aluminium. *Appl. Microbiol. Biotechnol.*, 52: 787- 790.

Johnson, R. and Bardal, E. (1985). Cathodic properties of different stainless steels in natural seawater. *Corrosion* 41: 296- 309.

Juzeliunas, E., Ramanauskas, R., Lugauskas, A., Leinartas, K., Samulevicien, M. and Sudavicius, A. (2006). Influence of wild strain *Bacillus* metals: from corrosion acceleration to environmentally friendly protection. *Electrochim. Acta.*, 51: 6085- 6090.

Kremer, M. L. (1985). Complex versus free radical mechanism for the catalytic decomposition of H_2O_2 by ferric ions. *Int. J. Chem. Kinet.*, 17: 1299.

References

L'Hostis, V., Dagbert, C. and Feron, D. (2003). Electrochemical behaviour of metallic materials used in seawater-interactions between glucose oxidase and passive layers. *Electrochim. Acta.*, 48: 1451- 1458.

Lai, M. E., Scotto, V. and Bergel, A. (1999). *Proceedings of the 10th International Congress on Marine Corrosion and Fouling,* The University of Melbourne, Melbourne, Australia.

Landoulsi, J., Dagbert, C., Richard, C., Sabot, R., Jeannin, M., El-Kirat, K. and Pulvine, S. (2009). Enzyme-induced ennoblement of AISI 316 L stainless steel: focus on pitting corrosion behaviour. *Electrochim. Acta.*, 54: 7401- 7406.

Landoulsi, J., Kirat, K. E., Richard, C., Sabot, R., Jeannin, M. and Pulvin, S. (2008). Glucose oxidase immobilization on stainless steel to mimic the aerobic activities of natural biofilms. *Electrochim. Acta.*, 54: 133- 139.

Lappin-Scott, H. M. and Costerton, J. W. (1995). In *Microbial Biofilms*; Lappin-Scott, H. M., Costerton, J. W., Eds. Cambridge University Press: Cambridge, U.K.

LeBozec, N., Compere, C., L'Her, M., Laoenan, A., Costa, D. and Marcus, P. (2001). Influence of stainless-steel surface treatment on the oxygen reduction reaction in seawater. *Corros. Sci.*, 43: 765.

LeBozec, N., L'Her, M., Laovenan, A., Costa, D., Marcus, P. and Compere, C. (1998). Ageing of passivated materials in seawater: Study of the oxygen reduction reaction. *Proceedings of Euromat,* Lisbon, Portugal.

Lee, A. K. and Newman, D. K. (2003). Microbial iron respiration: impacts on corrosion processes. *Appl. Microbiol. Biotechnol.* 62: 134- 139.

Lewandowski, Z., Lee, W. C., Characklis, G. and Little, B. (1989). Dissolved oxygen and pH microelectrode measurements at water immersed metal surfaces. *Corrosion.*, 45: 92.

Little, B. J., Pope, R. K., Daulton, T. and Ray, R. I. (2001). Application of transmission electron microscopy to microbiologically influenced corrosion. *Corrosion/2001*. NACE International, Huston, TX., 01266.

Little, B. J., Wagner, P., Hart, K., Ray, R., Lavoie, D., Nealson, K. and Aguilar, C. (1997). The role of metal-reducing bacteria in microbiologically influenced corrosion. In: *Proceeding NACE Corrosion*. NACE International, Houston, TX., 97: 215.

Lovley, D. R., Holmes, D. E. and Nevin, K. P. (2004). Dissimilatory Fe (III) and Mn (IV) reduction. *Adv. Microb. Physiol.*, 49: 219- 286.

Marconnet, C., Dagbert, C., Roy, M. and Feron, D. (2008). Stainless steel ennoblement in freshwater: from exposure tests to mechanisms. *Corrosion Sci.*, 50: 2342- 2352.

Maruthamuthu, S. R., Sathianarayanan, G., Angappan, S., Eashwar, S. M. and Balakrishnan, K. (1996). Contributions of oxide film and bacterial metabolism to the ennoblement process: evidence for a novel mechanism. *Curr. Sci.*, 71: 315- 320.

Mehanna, M., Basseguy, R., Delia, M. L., Laurence, G., Marie, D., and Bergel, A. (2008). New hypotheses for hydrogenase implication in the corrosion of mild steel. *Electrochim. Acta* 54: 140- 147.

Migahed, M. A. and Al-Sabagh, A. M. (2009). Beneficial role of surfactant as corrosion inhibitor in petroleum industry. *Chem. Eng. Commun.*, 196: 1054- 1075.

Miranda, E., Bettencourt, M., Botana, F., Cano, M., Sánchez-Amaya, J. and Corzo, A. (2006). Biocorrosion of carbon steel alloys by an hydrogenotrophic sulfate-reducing bacterium *Desulfovibriocapillatus* isolated from a Mexican oil field separator. *Corrosion Sci.*, 48: 2417- 2431.

Mobin, M., Khan, M. A. and Parveen, M. (2011). Inhibition of mild steel corrosion in acidic medium using starch and surfactants additives. *J. Appl. Polym. Sci.*, 121: 1558- 1565.

Mobin, M., Zehra, S. and Aslam, R. (2016). L-Phenylalanine methyl ester hydrochloride as a green corrosion inhibitor for mild steel in hydrochloric acid solution and the effect of surfactant additive. *RSC Adv.*, 6: 5890- 5902.

Mollica, A. (1992). Biofilm and corrosion on active-passive alloys in seawater. *Int. Biodeterior. Biodegrad.*, 29: 213- 229.

Mollica, A., Travers, E. and Ventura, G. (1990). 11th ZntlCorros Cong. Florence., IV: 1- 15.

Morad, M. S. (2005). Effect of amino acids containing sulfur on the corrosion of mild steel in phosphoric acid solutions containing Cl-, F- and Fe^{3+} ions : behaviour under polarization conditions. *J. Appl. Electrochem.*, 35: 889- 895.

Murphy, M. G. and Condon, S. (1984). Correlation of oxygen utilization and hydrogen peroxide accumulation with oxygen induced enzymes in Lactobacillus plantarum cultures. *Arch. Microbiol.*, 138: 44.

Naclerio, G., Baccigalupi, L., Caruso, C., de Felice, M. and Ricca, E. (1995). *Bacillus subtilis* vegetative catalase is an extracellular enzyme. *Appl. Environ. Microbiol.* 61: 4471- 4473.

Nealson, K.H. and Saffarini, D. (1994). Iron and manganese in anaerobic respiration: environmental significance, physiology, and regulation. *Annu. Rev. Microbiol.*, 48: 311- 343.

Okuyama, M. and Haruyama, S. (1990). The cathodic reduction of oxygen on stainless steels in a neutral solution. *Corros. Sci.*, 31: 521.

Patel, A. K., Singhania, R. R. and Pandey, A. (2017). Production, Purification, and Application of Microbial Enzymes. Production, Biocatalysis and Industrial Applications. *Biotechnology of Microbial Enzymes.*, 13-41. https://doi.org/10.1016/B978-0-12-803725-6.00002-9.

Prochnow, A. M., Lucas-Elio, P., Egan, S., Thomas, T., Webb, J. S., Sanchez-Amat, A. and Kjelleberg, S. (2008). Hydrogen peroxide linked to lysine oxidase activity facilitates biofilm differentiation and dispersal in several gram-negative bacteria. *J. Bacteriol.* 190: 5493- 5501.

Rusling, J. F. (1998). Enzyme bioelectrochemistry in cast bio-membrane-like films. *Acc. Chem. Res.,* 31: 363- 369.

Salvago, G. and Magagnin, L. (2001). Biofilm effect on the cathodic and anodic processes on stainless steel in seawater near the corrosion potential: Part 2-Oxygen reduction on passive metal. *Corrosion.*, 57: 759.

Scotto, V., Di-Cintio, R. and Marcenaro, G. (1985). The influence of marine aerobic microbial film on stainless steel corrosion behaviour. *Corrosion Sci.* 25: 185- 194.

Shams El Din, A. M., Saber, T. M. H. and Hammoud, A. A. (1996). Biofilm formation on stainless steels in Arabian Gulf water. *Desalination.* 107: 251.

Shi, X., Avci, R., Geiser, M. and Lewandowski, Z. (2003). Comparative study in chemistry of microbially and electrochemically induced pitting of 316 L stainless steel. *Corrosion Sci.*, 45: 2577- 2595.

Sutherland, I. W. (2001). The biofilm matrix - An immobilized but dynamic microbial environment. *Trends Microbiol.*, 9: 222.

Tilman, D. (1982). *Resource Competition and Community Structure*. Princeton University Press, Princeton.

Washizu, N., Hiroyuki, M. and Toshiaki, K. (2001). Relation between ennoblement of open-circuit potentials for SUS 316L induced by marine microorganisms and hydrogen peroxide in biofilms. *Corros. Eng.*, 50: 330.

Washizu, N., Katada, Y. and Kodama, T. (2004). Role of H_2O_2 in microbially influenced ennoblement of open circuit potentials for type 316 L stainless steel in seawater. *Corrosion Sci.* 46: 1291- 1300.

Xu, K., Dexter, S. C. and Luther, G. W. (1998). Voltametric microelectrodes for biocorrosion studies. *Corrosion.*, 54: 814.

Yang, H. H. and McCreery, R. L. (2000). Elucidation of the mechanism of dioxygen reduction on metal-free carbon electrodes. *J. Electrochem. Soc.*, 147: 3420.

Yeager, E. (1986). Dioxygen electrocatalysis: mechanism in relation to catalyst structure. *J. Mol. Catal.*, 38: 5.

Zepp, R. G., Faust, B. C. and Hoignè, J. (1992). Hydroxyl radical formation in aqueous reactions (pH 3–8) of iron (II) with hydrogen peroxide: the photo-Fenton reaction. *Environ. Sci. Technol.*, 26: 313.

Zika, R. G., Saltzman, E. S. and Cooper, W. J. (1985). Hydrogen peroxide concentration in the Peru upwelling area. *Mar. Chem.*, 17: 265.

Chapter 7

Beech, I. B. and Gaylarde, C. C. (1999). Recent advances in the study of biocorrosion: An overview. *Revista de Microbiologia.*, 30: 177- 190.

Eder, K. (1995). Gas chromatographic analysis of fatty acid methyl esters. *J Chromatogr B Biomed Appl.*, 15, 671(1-2): 113-31. doi: 10.1016/0378-4347(95)00142-6.

Karn, S. K., Bhambri, A., Jenkinson, I. R., Duan, Z. and Kumar, A. (2020). The role of biomolecules in corrosion induction and inhibition. *Corrosion Review.*, 38(5): 1- 18. doi: 10.1515/corrrev-2019-0111.

Karn, S. K. and Kumar, A. (2019). Sludge, Next Paradigm for Enzyme Extraction and Energy Generation. *Preparative Biochemistry & Biotechnology.*, 49 (2): 105-116, 201-205. https://doi.org/10.1080/10826068.2019.1566146.

Kitaguchi, H., Ohkubo, K., Ogo, S. and Fukuzumi, S. (2006). Electron-transfer oxidation properties of unsaturated fatty acids and mechanistic insight into lipoxygenases. *J. Phys. Chem.*, 110: 1718- 1725.

Klinman, J. P. (2007). Linking protein dynamics to function. *Faseb. J.*, 21: A645.

Nicholls, D. G. and Ferguson, S. J. (2002). *Bioenergetics 3* (Academic Press, London, 2002). *Biochem.* (Moscow). 69: 818- 819.

Peterson, D. A. (1991). Enhanced electron transfer by unsaturated fatty acids and superoxide dismutase. *Free Radic. Res.,* 12: 161- 166.

Romanowicz, L., Jaworski, S., Galewska, Z. and Gogiel, T. (2008). Separation and Determination of Fatty Acids from Lipid Fractions by High-Performance Liquid Chromatography: Cholesterol Esters of Umbilical Cord Arteries. *Toxicology Mechanisms and Methods.,* 18: 509- 513. doi: 10.1080/15376510701623912.

Chapter 8

Andersen, K., Bird, K. L., Rasmussen, M., Halie, J., Breunning-Madsen, H., Kajer, K. H., Orlando, L., Gilbert, M. T. P. and Willerslev, E. (2011). Meta-of "dirt" DNA from soil reflects vertebrate biodiversity. *Mol. Ecol.,* 21: 1966–1979.

Blum, S. A. E., Lorenz, M. G. and Wackernagel, W. (1997). Mechanism of retarded DNA degradation and prokaryotic origin of DNases in nonsterile soils. *Syst. Appl. Microbiol.,* 20: 513–521.

Bockelmann, U., Janke, A. and Kuhn, R. (2006). Bacterial extracellular DNA forming a defined network like structure. *FEMS Microbiol. Lett.* 262: 31–38.

Bunce, M., Szulkin, M., Lerner, H. R. L., Barnes, I., Shapiro, B., Cooper, A. and Holdaway, R. N. (2005). Ancient DNA provides new insights into the evolutionary history of New Zealand's extinct giant eagle. *PLoS Biol.,* 3: e9.

Crecchio, C. and Stotzky, G. (1998). Binding of DNA on humic acids: effect on transformation of *Bacillus subtilis* and resistance to DNase. *Soil Biol. Biochem.,* 30: 1061–1067.

Dejean, T., Valentini, A., Duparc, A., Pellier-Cuit, S., Pompanon, F., Taberlet, P. and Miaud, C. (2011). Persistence of environmental DNA in freshwater ecosystems. *PLoS ONE.,* 6(8): e23398. doi:10.1371/journal. pone,0023398.

Flaviani, F., Schroeder, D. C., Balestreri, C., Schroeder, J. L., Moore, K., Paszkiewicz, K., Pfaff, M. C. and Rybicki, P. E. (2017). Apelagic microbiome (viruses to protists) from a small cup of seawater. *Viruses.,* 9: 47.

Haile, J., Holdaway, R., Oliver, K., Bunce, M., Gilbert, M. T., Nielsen, R., Munch, K., Ho, S. Y., Shapiro, B. and Willerslev, E. (2007). Ancient DNA chronology within sediment deposits: are paleobiological reconstructions possible and is DNA leaching a factor? *Mol. Biol. Evol.,* 24: 982–989.

Hebsgaard, M. B., Gilbert, M. T. P., Arneborg, J., Heyn, P., Allentoft, M. E., Bunce, M., Munch, K., Schweger, C. and Willerslev, E. (2009). The farm beneath the sand–an archaeological case study on ancient "dirt" DNA. *Antiquity.,* 83: 430–444.

Higuchi, R., Von-Beroldingen, C. H., Sensabaugh, G. F. and Erlich, H. A. (1988). DNA typing from single hairs. *Nature.,* 332: 543–546.

Hofreiter, M., Mead, J. I., Martin, P. and Poinar, H. N. (2003). Molecular caving. *Curr. Biol.,* 13: 693–695.

Huang, Y. T., Lowe, D. J., Churchman, G. J. and Schipper, L. A. (2014). Carbon storage and DNA adsorption in allophanic soils and paleosols. In: *Soil Carbon,* McSweeney, K. and Hartemink, A. E. (Eds.). Springer International, New York., 163–172.

Jenkins, D. L., Davis, L. G., Stafford, T. W., Campos, P. F., Hockett, B. and Jones, G. T. (2012). Clovis age Western stemmed projectile points and human coprolites at the Paisley Caves. *Science.*, 337: 223–228.

Karn, S. K., Bhambri, A., Jenkinson, I. R., Duan, J. and Kumar, A. (2020). The roles of biomolecules in corrosion induction and inhibition of corrosion: a possible insight. *Corros. Rev.*, 1–19. https://doi.org/10.1515/corrrev-2019-0111.

Lacoursiere Roussel, A., Kimberly, H., Normandeau, E., Grey, E. K., Archambault, D. K., Lodge, D. M., Leduc, N. and Bernatchez, L. (2018). eDNA metabarcoding as a new surveillance approach for coastal Arctic biodiversity. *Ecol. Evol.*, 16: 7763–7777.

Lauw, Y., Horne, M. D., Rodopoulos, T., Nelson, A. and Leermakers, F. A. M. (2010). Electrical double-layer capacitance in room temperature ionic liquids: ion-size and specific adsorption effects. *J. Phys. Chem. B.*, 114: 11149–11154.

Levy-Booth, D. J., Campbell, R. G. and Gulden, R. H. (2007). Cycling of extracellular DNA in the soil environment. *Soil Biol. Biochem.*, 39: 2977–2991.

Liu, H., Steigerwald, M. L. and Nuckolls, C. (2009). Electrical double layer catalyzed wet-etching of silicon dioxide. *J. Am. Chem. Soc.*, 131: 17034–17035.

Lorenz, M. G. and Wackernagel, W. (1987). Adsorption of DNA to sand and variable degradation rates of adsorbed DNA. *Appl. Environ. Microbiol.*, 53: 2948–2952.

Meier, P. and Wackernagel, W. (2003). Mechanisms of homology facilitated illegitimate recombination for foreign DNA acquisition in transformable *Pseudomonas stutzeri*. *Mol. Microbiol.*, 48: 1107–1118.

Nichols, R. V., Nigsson, H. K., Danell, K. and Spong, G. (2012). Browsed twig environmental DNA: diagnostic PCR to identify ungulate species. *Mol. Ecol. Resour.*, 12: 983–989.

Nielsen, K. M., Johnsen, P. J., Bensasson, D. and Daffonchio, D. (2007). Release and persistence of extracellular DNA in the environment. *Environ. Biosaf. Res.*, 6: 37–53.

Pietramellara, G., Ascher, J., Ceccherini, M. T., Nannipieri, P. and Wenderoth, D. (2007). Adsorption of pure and dirty bacterial DNA on clay minerals and their transformation frequency. *Biol. Fertil. Soils.*, 43: 731–739.

Poinar, H. N. (1998). Molecular: dung and diet of the extinct ground sloth shastensis. *Science.*, 281: 402–406.

Qin, Z., Ou, Y. and Yang, L. (2007). Role of autolysin mediated DNA release in biofilm formation of Staphylococcus epidermidis. *Microbiology.*, 153: 2083–2092.

Steinberger, R. E. and Holden, P. A. (2005). Extracellular DNA in single and multiple species unsaturated biofilm. *Appl. Environ. Microbial.* 71: 5404–5410.

Steinem, C., Janshoff, A., Lin, V. S. Y., Vo, N. H. and Ghadiri, M. R. (2004). DNA hybridization-enhanced porous silicon corrosion: mechanistic investigations and prospect for optical interferometric biosensing. *Tetrahedron.*, 60: 11259–11267.

Strausberger, B. M. and Ashley, M. V. (2001). Eggs yield nuclear DNA from egg-laying female cowbirds, their embryos and offspring. *Conserv. Genet.*, 2: 385–390.

Taberlet, P. and Bouvet, J. (1991). A single plucked feather as a source of DNA for bird genetic-studies. *Auk.*, 108: 959–960.

Taberlet, P. and Fumagalli, L. (1996). Owl pellets as a source of DNA for genetic studies of small mammals. *Mol. Ecol.*, 5: 301–305.

Taberlet, P., Mattock, H., Dubois-Paganon, C. and Bouvet, J. (1993). Sexing free-ranging brown bears using hairs found in the field. *Mol. Ecol.*, 2: 399–403.

Trevors, J. T. (1996). Nucleic acids in the environment. *Curr. Opin. Biotechnol.*, 7: 331–336.

Valiere, N. and Taberlet, P. (2000). Urine collected in the field as a source of DNA for species and individual identification. *Mol. Ecol.*, 9: 2150–2152.

Watanabe, M., Sasaki, K., Nakashimada, Y., Kakizono, T., Noparatnaraporn, N. and Nishio, N. (1995). Growth and flocculation of marine photosynthetic bacterium Rhodovulum sp. *Appl. Microbiol. Biotechnol.*, 50: 682–691.

Zhao, Y., Lawrie, J. L., Beavers, K. R., Laibinis, P. E. and Weiss, S. M. (2014). Effect of DNA-induced corrosion on passivated porous silicon biosensors ACS. *Appl. Mater. Interfaces.*, 6: 13510–13519.

Chapter 9

Cragnolino, G. and Tuovinen, O. H. (1984). *Internal. Biodeterior.*, 20: 9- 26.

Dubois, M., Gilles, K. A., Hamilton, J. K., Rebers, P. A. and Smith, F. (1956). Colorimetric method for determination of sugars and related substances. *Anal. Chem.*, 28(3): 350- 356.

Kapley, A., Lampel, K. and Purohit, H. J. (2001). Rapid detection of Salmonella in water samples by multiplex PCR. *Water Environ. Res.* 73: 461–465.

Karn, S. K., Bhambri, A., Jenkinson, I. R., Duan, J. and Kumar, A. (2020). The roles of biomolecules in corrosion induction and inhibition of corrosion: a possible insight. *Corros. Rev.*, 1- 19. https://doi.org/10.1515/corrrev-2019-0111.

Karn, S. K., Chakrabarti, S. K. and Reddy, M. S. (2010). Isolation and characterization of pentachlorophenol degrading *Bacillus* sp. Isolated from secondary sludge of pulp and paper industry. *International Biodeterioration and Biodegradation.* 64: 609-613. ISSN: 0964- 8305.

Karn, S. K., Chakrabarti, S. K. and Reddy, M. S. (2011). Degradation of penta- chlorophenol by *Kocuria*sp. CL2 isolated from secondary sludge of pulp and paper mill. *Biodegradation.* 22(1): 63- 69. ISSN: 0923- 9820.

Lawrence, J. R., Swerhone, G. D. W., Kuhlicke, U. and Neu, T. R. (2007). In-situ evidence for microdomains in the polymer matrix of bacterial microcolonies. *Can. J. Microbiol.* 53: 450–458.

LeChevallier, M. W., Seidler, R. J. and Evans, T. M. (1980). Enumeration and characterization of standard plate count bacteria in chlorinated and raw water supplies. *Appl. Environ. Microbiol.*, 40: 922- 930.

Miller, J. D. A. (1981). *Economic Microbiology*, ed. Rose, A. H. Acad. Press, London: Academic., 1: 150- 202.

Neu, T. R. and Lawrence, J. R. (1999). In *Microbial Extracellular Polymeric Substances* eds Wingender, J., Neu, T. and Flemming, H. C. Springer, Heidelberg. 22- 47.

Pope, D. H., Duquette, D., Wayner, P. C. and Johannes, A. H. (1984). Microbiologically Influenced Corrosion: A State-of-the-Art Review. Columbus, OH: Materials Techno! *Insl. Chem. Proc. Ind.*, 1- 76.

Reasoner, D. J. and Geldreich, E. E. (1985). A new medium for the enumeration and subculture of bacteria from potable water. *Appl. Environ. Microbiol.* 49: 1- 7.

Tiller, A. K. (1982). Corrosion Processes, ed. Parkins, R. N. London: *Appl. Science Pub.*, 115- 59.

Wagner, M., Ivleva, N. P., Haisch, C., Niessner, R. and Horn, H. (2009). Combined use of Confocal Laser Scanning Microscopy (CLSM) and Raman Microscopy (RM): investigations on EPS-matrix. *Water Res.* 43: 63–76.

Weisburg, W. G., Barns, S. M., Pelletier, D. A. and Lane, D. J. (1991). 16S ribosomal DNA amplification for phylogenetic study. *J. Bacteriol.* 173: 697–703.

Wingender, J. and Jaeger, K. E. (2002). In *Encyclopedia of Environmental Microbiology* ed. Bitton, G. Wiley, New York., 1207- 1223.

Wrangstadh, M., Szewzyk, U., Östling, J. and Kjelleberg, S. (1990). Starvation-specific formation of peripheral exopolysaccharide by a marine *Pseudomonas* sp., strain S9. *Appl. Environ. Microbiol.* 56: 2065–2072.

Chapter 10

Bailey, J. E. (1991). Toward a science of metabolic engineering. *Science.*, 252: 1668-1675.

Desbrosses, G. G., Kopka, J. and Udvardi, M. K. (2005). *Lotus japonicus* metabolic profiling. Development of gas chromatography-mass spectrometry resources for the study of plant-microbe interactions. *Plant Physiol.*, 137: 1302- 1318.

Fiehn, O. (2002). Metabolomics - the link between genotypes and phenotypes. *Plant Mol Biol.*, 48: 155- 171.

Frondi, M. and Lio, P. (2015). Multi-omics and metabolic modelling pipelines: challenges and tools for systems microbiology. *Microbiological Research.* 171: 52, 64.

Gehlenborg, N., O'Donoghue, S. I., Baliga, N. S., Goesmann, A., Hibbs, M. A., Kitano, H., Kohlbacher, O., Neuweger, H., Schneider, R., Tenenbaum, D., & Gavin, A. C. (2010). Visualization of omics data for systems biology. *Nature Methods.* 7: S56-S68.

Gieger, C. L., Geistlinger, E., Altmaier, M., Hrabé de Angelis, F., Kronenberg T., Meitinger, Mewes, H. W. and Wichmann, H. E. (2008). Genetics meets metabolomics: A genome-wide association study of metabolite profiles in human serum. *PLoS Genet.*, *4*: e1000282.

Kato, H., Takahashi, S. and Saito., K. (2011). Omics and integrated omics for the promotion of food and nutrition science. *Journal of Traditional and Complementary Medicine.* 1: 25-30.

Kell, D. B., Brown, M. and Davey, H. M. (2005). Metabolic footprinting and systems biology: the medium is the message. *Nat Rev Microbiol.*, 3: 557- 565.

Misra, B. B. (2018). New tools and resources in metabolomics: 2017-2017. *Electrophoresis.* 39: 909- 923.

Misra, B. B., Fahramann, J. F. and Grapov, D. (2017). Review of emerging metabolomic tools and resources: 2015-2016. *Electrophoresis.* 38: 2257- 2274.

Misra, B. B. and van der Hooft, J. J. (2016). Updates in metabolomics tools and resources: 2014-2015. *Electrophoresis.* 37: 86-110.

Muller, E. E. L., Pinel, N., Laczny, C. C., Hoopmann, M. R., Narayanasamy, S., Lebrun, L. A., Roume, H., Lin, J., May, P., Hicks, N. D., Heintz-Buschart, A., Wampach, L., Liu, C. M., Price, L. B., Gillece, J. D., Guignard, C., Schupp, J. M., Vlassis, N., Baliga, N. S., Wilmes, P. (2014). Community-integrated omics links dominance of a microbial generalist to fine-tuned resource usage. *Nature Communications.,* 5: 5603.

Nielsen, J. (2001). Metabolic engineering. *Appl. Microbiol. Biotechnol.,* 55: 263-283.

Oliver, S. G., Winson, M. K., Kell, D. B. and Baganz, F. (1998). Systematic functional analysis of the yeast genome. *Trends Biotechnol.,* 16: 373- 378.

Oveland, E., Muth, T., Rapp, E., Martens, I., Berven, F. S. and Barnes, H. (2015). Viewing the proteome: how to visualize proteomics data? *Proteomics.,* 15: 134-1355.

Pavlopoulos, G. A., Malliarakis, D., Papanikolaou, N., Theodosiou, T., Enright, A. J. and Iliopoulos, I. (2015). Visualizing genome and systems biology: technologies, tools, implementation techniques and trends, past, present and future. *GigaScience.,* 4: 1-27.

Ritchie, M. D., Holzinger, E. R., Li, R., Pendergrass, S. A. and Kim, D. (2015). Methods of integrating data to uncover genotype-phenotype interactions. *Nature Reviews Genetics.,* 16: 85- 97.

Roessner, U., Wagner, C., Kopka, J., Trethewey, R. N. and Willmitzer, L. (2003). Simultaneous analysis of metabolites in potato by gas chromatography-mass spectrometry. *Plant J.,* 23: 131- 142.

Villas-Bôas, S. G., Noel, S. and Lange, G. A. (2006). Extracellular metabolomics: a metabolic footprinting approach to assess fiber degradation in complex media. *Anal Biochem.,* 349: 297- 305.

Villas-Bôas, S. G., Roessner, U., Hansen, M., Smedsgaard, J. and Nielsen, J. (2007). *Metabolome Analysis: An Introduction.* John Wiley & Sons, Inc., Hoboken, NJ.

Index

A

access, 16, 46, 54, 86
acid, 8, 27, 28, 32, 42, 43, 45, 52, 54, 55, 58, 63, 65, 66, 73, 74, 76, 82, 118, 119, 121, 122
acidic, 2, 15, 27, 51, 53, 55, 58, 85, 121
adhesion, 15, 16, 20, 49, 50, 86, 103, 104, 105
age, 51, 83, 124
aggregation, 49, 50, 52, 116
algae, 15, 23, 25, 26, 51, 53
aluminium, 1, 3, 119
amino acids, 55, 57, 58, 65, 66, 72, 117, 119, 121
ammonia, 40, 107, 110
antibacterial coatings, ix, 1, 3
antibacterial stainless steels, ix, 1, 3
assessment, 40, 45, 111
ATP, 29, 71, 73, 93
attachment, x, 8, 13, 14, 15, 16, 17, 18, 20, 26, 35, 46, 49, 103, 105
automated Intergenic Spacer Analysis (ARISA), 31, 38

B

bacteria, ix, 1, 2, 3, 4, 5, 7, 8, 9, 10, 11, 12, 13, 14, 15, 16, 17, 21, 23, 24, 25, 26, 27, 28, 29, 31, 36, 37, 38, 39, 41, 45, 50, 52, 53, 54, 55, 57, 58, 62, 64, 65, 68, 69, 82, 83, 85, 87, 89, 97, 98, 99, 100, 101, 102, 104, 105, 107, 109, 110, 113, 114, 116,117, 118, 119, 120, 121, 125, 126
bacterial cells, 8, 21, 50, 93

bacterium, 9, 11, 15, 26, 28, 53, 55, 56, 62, 63, 82, 97, 101, 111, 113, 117, 118, 121, 125
base, 35, 38, 39, 41, 55, 85, 109, 110
biocatalysts, 10, 92, 95
biochemical characterization, x, 23, 31, 89
biocide enhancers, ix, 1, 3
biocorrosion, vii, ix, 1, 2, 4, 5, 11, 12, 49, 53, 59, 62, 63, 66, 71, 85, 91, 94, 97, 98, 101, 102, 104, 113, 116, 117, 118, 119, 121, 122
biodiversity, 31, 32, 45, 47, 62, 80, 82, 106, 108, 123, 124
biofilm, v, vii, ix, x, 1, 2, 5, 6, 7, 9, 10, 11, 13, 14, 15, 16, 17, 18, 19, 20, 21, 27, 49, 50, 51, 53, 54, 55, 59, 60, 61, 62, 63, 64, 66, 67, 68, 69, 71, 73, 74, 82, 85, 86, 87, 88, 89, 91, 92, 95, 97, 98, 99, 100, 101, 102, 103, 104, 105, 113, 114, 115, 116,117, 118, 119, 120, 121, 122, 124
biofilm matrix, ix, x, 16, 49, 54, 55, 62, 71, 82, 103, 114, 116, 122
biomolecules, vii, ix, 56, 58, 71, 85, 86, 92, 104, 106, 117, 122, 124, 125
bioremediation, 27, 91, 109, 147
biotechnology, 46, 94, 103, 104, 113, 147
breakdown, 54, 58, 60

C

calcium, 51, 88, 90
carbohydrates, vii, 32, 50, 52, 54, 55, 72, 85, 88

carbon, 1, 3, 26, 27, 28, 50, 51, 52, 54, 58, 62, 63, 64, 65, 88, 89, 93, 99, 101, 105, 106, 113, 114, 119, 121, 122
carbon steel, 1, 3, 26, 63, 89, 99, 101, 105, 106, 113, 114, 119, 121
carboxyl, 49, 54, 55, 58
catalysis, 10, 11, 58, 63, 68
cell attachment, x, 49, 105
cells, x, 5, 6, 8, 9, 10, 13, 14, 15, 16, 18, 19, 20, 21, 24, 26, 32, 33, 46, 47, 49, 50, 52, 53, 57, 58, 59, 62, 63, 64, 73, 82, 83, 85, 86, 90, 91, 92, 93, 94, 95, 96, 104, 105
challenges, 2, 92, 126
charge density, x, 79, 81
chemical, ix, 1, 2, 11, 14, 17, 20, 28, 48, 51, 52, 56, 57, 58, 59, 60, 61, 67, 71, 75, 77, 79, 85, 91, 94, 95, 96, 98, 108
chemical properties, 14, 17, 51, 62, 96
chromatography, 71, 75, 77, 92, 94, 126, 127
chromium, 88, 104, 106
classification, 32, 45, 67, 95, 111
cloning, 39, 45, 111
CO^2, 11, 15, 28, 95
coding, 35, 38, 91, 110
colonization, 17, 19, 97
community, 7, 13, 16, 21, 24, 33, 35, 36, 38, 39, 40, 41, 42, 43, 44, 45, 47, 54, 64, 65, 99, 105, 108, 109, 110, 111, 112
competition, 21, 69, 145
complexity, 19, 39, 43, 93
composition, 8, 16, 33, 39, 43, 44, 48, 50, 51, 52, 56, 64, 66, 76, 87, 105, 108
compost, 35, 38, 40, 41, 43, 45, 110, 111
composting, 35, 36, 39, 40, 108, 109, 110, 111, 112
compounds, 2, 7, 9, 16, 18, 28, 46, 50, 54, 59, 63, 69, 72, 83, 94, 95
confocal laser scanning microscopy (CLSM), vii, 44, 85, 87, 126
connectivity, 91, 92, 95
consumption, 4, 7, 59, 67
contamination, 14, 46, 102, 111

copper, 15, 16, 56, 66, 97, 98, 100, 103, 111, 119
corrosion, ix, x, 1, 2, 3, 4, 5, 6, 7, 8, 9, 10, 11, 12, 15, 16, 18, 23, 25, 27, 49, 50, 54, 55, 57, 59, 60, 61, 62, 63, 64, 65, 66, 71, 72, 73, 79, 80, 81, 82, 83, 85, 86, 89, 96, 97, 98, 99, 100, 101, 102, 103, 104, 105, 106, 113, 114, 115, 117, 118, 119, 120, 121, 122, 124, 125
cost, 12, 46, 52, 80
crude oil, 5, 11, 54, 109
cultivation, 108, 113, 147
culture, 31, 32, 33, 42, 52, 60, 64, 87, 89, 95, 96, 112
cycles, 27, 28, 90

D

damages, 3, 5, 11, 12
database, 48, 73, 96
decomposition, 9, 26, 69, 119
degradation, 1, 3, 8, 9, 15, 63, 68, 106, 109, 111, 123, 124, 127
denaturation, 38, 39, 41, 90
depolarization, 1, 5, 7, 63, 98
deposition, 9, 17, 26, 60, 69
deposits, 7, 9, 26, 86, 123
detection, 35, 39, 79, 80, 84, 87, 92, 125
dispersion, 13, 18, 20
distribution, 3, 14, 26, 36, 37, 43, 62, 63, 67, 80, 87, 95, 98, 101, 104, 111, 117, 118
diversity, v, vi, ix, x, 19, 23, 24, 31, 32, 33, 34, 35, 36, 37, 38, 39, 40, 41, 42, 43, 44, 45, 47, 48, 52, 62, 80, 96, 101, 106, 107, 108, 109, 110, 111, 112, 114, 115
DNA, vi, ix, 16, 21, 24, 31, 34, 35, 36, 37, 39, 40, 41, 42, 43, 44, 45, 46, 47, 49, 79, 80, 81, 82, 83, 84, 87, 89, 90, 93, 108, 109, 110, 111, 112, 123, 124, 125, 126
donors, 27, 64, 65, 100
drinking water, 1, 7, 14, 98, 104, 105

E

E. coli, 15, 37, 90, 108

ecology, 43, 45, 47, 98, 103, 107, 109, 110, 111, 112
ecosystem, 18, 33, 38, 40, 79, 80
eDNA, vi, x, 13, 79, 80, 81, 82, 83, 124
electrolyte, 2, 9, 60, 81
electron, ix, 1, 2, 6, 7, 10, 11, 27, 28, 55, 59, 64, 65, 67, 68, 71, 72, 73, 74, 85, 107, 118, 123
electron transfer, ix, 2, 55, 59, 64, 65, 71, 73, 74, 118, 123
electrophoresis, 34, 35, 39, 40, 41, 110, 111
energy, 2, 26, 27, 56, 85
engineering, ix, 92, 93, 95, 126, 127
environment, vi, ix, 1, 2, 3, 5, 7, 8, 11, 13, 14, 15, 16, 17, 18, 19, 21, 23, 24, 27, 28, 31, 33, 34, 36, 39, 42, 43, 44, 45, 46, 49, 50, 52, 56, 57, 59, 62, 64, 79, 80, 82, 83, 85, 89, 91, 92, 94, 95, 96, 97, 101, 102, 104, 105, 106, 108, 109, 112, 113, 116, 117, 121, 122, 123, 124, 125, 126, 145, 147
environmental conditions, 31, 35, 49, 52, 56, 62, 79
enzymatic activity, 59, 62, 68, 95
enzymes, vi, ix, x, 1, 2, 5, 16, 35, 36, 46, 50, 57, 58, 59, 60, 61, 62, 64, 65, 66, 67, 68, 69, 73, 82, 83, 87, 95, 115, 118, 119, 120, 121, 122
EPS, vi, ix, x, 8, 14, 17, 49, 50, 51, 52, 53, 54, 55, 56, 69, 87, 113, 114, 115, 126
equilibrium, 6, 10, 58
equipment, 2, 3, 18, 41, 46
ester, 55, 73, 121
eukaryotic, 38, 40, 108
evidence, 80, 116, 120, 125
evolution, 23, 31, 110, 112, 119
exopolysaccharide (EPS), vi, ix, x, 8, 14, 17, 49, 50, 51, 52, 53, 54, 55, 56, 69, 87, 113, 114, 115, 126
exopolysaccharides, 52, 87, 105, 114, 115, 116
exposure, 12, 21, 61, 80, 89, 120

extracellular polymeric substances, 3, 8, 10, 15, 16, 20, 21, 49, 50, 51, 52, 53, 54, 82, 86, 87, 113, 115, 116
extraction, 33, 35, 39, 45, 46, 75, 90, 92, 95, 147

F

factories, 18, 92, 95
fatty acids, 64, 65, 73, 74, 76, 108, 122, 123
fermentation, 32, 64, 65, 85, 89
films, ix, 62, 66, 67, 85, 121
fingerprinting, vii, 33, 38, 93, 94, 111
flagellum, 14, 15, 26, 103
fluorescence, 38, 47, 56, 75
food, 7, 13, 18, 19, 21, 51, 52, 54, 93, 100, 102, 103, 104, 105, 126
food industry, 18, 19, 100, 103
formation, ix, x, 6, 7, 9, 13, 14, 15, 17, 18, 19, 20, 21, 26, 42, 49, 51, 55, 57, 62, 63, 65, 67, 73, 82, 85, 88, 98, 99, 101, 102, 105, 107, 116, 121, 124, 126
fragments, 34, 35, 36, 38, 39, 40, 41, 44, 46, 108
freshwater, 24, 87, 99, 120, 123
fungi, 4, 17, 23, 25, 31, 35, 36, 41, 50, 53, 63, 111, 115

G

gas chromatography, 71, 75, 94, 126, 127
gel, 34, 35, 38, 39, 40, 41, 52, 90, 110, 111
genes, 19, 20, 24, 33, 34, 35, 39, 40, 44, 45, 110, 111, 112
genetics, 39, 96, 103
genome, 24, 31, 35, 36, 37, 42, 44, 91, 92, 107, 108, 109, 114, 126, 127
genomics, x, 79, 80, 91, 92, 94, 95, 96, 103
genotype, 93, 94, 127
Germany, 99, 103, 104, 116, 146
glucose, 57, 60, 85, 88, 89, 119, 120
growth, 8, 12, 13, 15, 17, 26, 27, 28, 49, 51, 52, 63, 69, 89, 93, 96, 107, 115

H

habitat, 14, 24, 44, 80, 114
heterogeneity, 24, 38, 39
human, 18, 20, 32, 97, 102, 110, 124, 126
hybridization, 31, 40, 42, 43, 44, 45, 82, 106, 108, 109, 124
hydrogen, 2, 5, 6, 7, 8, 9, 10, 11, 57, 60, 61, 63, 64, 67, 68, 85, 98, 118, 121, 122
hydrogen peroxide (H_2O_2), 57, 59, 60, 61, 64, 67, 68, 118, 121, 122
hydrogenase, 1, 5, 63, 64, 65, 101, 118, 120
hydroxide, 10, 62, 82

I

identification, 3, 24, 32, 33, 36, 39, 40, 42, 44, 45, 48, 85, 89, 92, 118, 125
India, xi, 145, 146
induction, 50, 61, 72, 81, 104, 106, 114, 122, 124, 125
industry, v, 2, 3, 7, 11, 13, 17, 18, 19, 77, 80, 100, 103, 106, 113, 117, 120, 125
inhibition, 55, 65, 66, 102, 104, 106, 114, 116, 117, 118, 119, 122, 124, 125
inhibitor, 119, 120, 121
initiation, 21, 116, 118
interface, 5, 8, 10, 15, 16, 17, 59, 60, 81, 98
ions, 7, 14, 15, 16, 17, 48, 54, 55, 73, 121
iron, ix, 1, 3, 5, 6, 7, 9, 10, 12, 26, 27, 29, 54, 55, 62, 63, 64, 65, 66, 73, 88, 99, 100, 107, 119, 120, 122
iron-oxidizing bacteria, ix, 5, 26, 27
iron-reducing bacteria, 7, 23, 27, 107
isolation, 33, 44, 108

K

kinetics, 7, 31, 43

L

Lactobacillus, 52, 53, 118, 121
lakes, 24, 27, 52
leaching, vii, 28, 83, 85, 88, 123
lead, 12, 62, 66, 67, 85, 94
lipases, 18, 57, 59, 62
lipids, vi, ix, x, 16, 47, 50, 52, 53, 71, 72, 73, 74, 75, 76, 77, 123
liquid chromatography, 71, 75, 92, 94
Listeria monocytogenes, 13, 19, 102, 103

M

macromolecules, x, 16, 49, 57, 62
magnesium, 1, 3, 51, 88
MALDI, 31, 48, 111
management, ix, 71, 110
manganese, 1, 3, 4, 28, 57, 60, 61, 69, 104, 107, 117, 121
mapping, 62, 67, 118
mass, 26, 31, 48, 55, 62, 75, 76, 86, 89, 93, 94, 95
mass spectrometry, 31, 48, 75, 76, 93, 94
material surface, x, 17, 18, 49
materials, 1, 3, 17, 35, 49, 51, 54, 63, 97, 103, 104, 110, 113, 117, 119, 120
matrix, ix, x, 5, 8, 13, 15, 16, 20, 31, 48, 49, 50, 51, 53, 54, 55, 56, 62, 67, 71, 73, 75, 82, 83, 86, 103, 113, 114, 116, 122
matrix-assisted laser desorption/ionization time-of-flight (MALDI-TOF), 31, 48, 111
matter, 63, 75, 103
measurements, 33, 80, 83, 89, 95, 118, 120
media, 60, 101, 127, 145
medical, 3, 13, 21, 102
metabolic, vii, 73, 94, 95, 126, 127
metabolism, 21, 23, 32, 58, 72, 73, 92, 93, 95, 101, 117, 119, 120
metabolites, ix, 2, 52, 53, 66, 72, 91, 92, 93, 94, 95, 96, 115, 126, 127
metabolome, 73, 91, 92, 96
metabolomics, vii, 91, 92, 93, 94, 95, 96, 110, 126, 127
metal, vii, ix, x, 1, 2, 5, 6, 7, 8, 9, 10, 15, 16, 17, 18, 28, 49, 50, 53, 54, 55, 56, 59, 60, 62, 63, 64, 65, 66, 67, 79, 85, 86, 88, 89, 95, 97, 98, 101, 102, 106, 113, 115, 116, 117, 119, 120, 121, 122

Index

metal ion, 1, 6, 8, 49, 54, 55, 63, 85, 97, 102, 117
microbes, 1, 2, 8, 10, 13, 14, 15, 16, 17, 19, 20, 23, 25, 27, 31, 32, 33, 35, 37, 38, 40, 45, 48, 49, 50, 52, 53, 65, 98, 100, 105, 110, 111, 115, 126, 146
microbial cells, 9, 10, 16, 19
microbial communities, 32, 35, 36, 38, 40, 41, 42, 43, 44, 64, 106, 108, 109, 110, 111
microbial diversity, ix, 23, 24, 31, 32, 34, 35, 36, 39, 44, 45, 80, 108, 110, 112
microbial-influenced corrosion (MIC), v, ix, 3, 10, 15, 49, 55, 67, 69, 98, 99, 101, 102, 118
microorganisms, v, ix, 2, 3, 4, 11, 13, 14, 15, 16, 17, 19, 23, 24, 25, 26, 27, 31, 32, 33, 35, 43, 45, 47, 48, 49, 51, 52, 53, 54, 55, 59, 60, 62, 64, 65, 66, 67, 68, 69, 71, 82, 83, 85, 86, 93, 102, 104, 107, 111, 115, 118, 122, 145
microscopy, 44, 85, 87, 97
mineralization, 91, 95, 96, 110
molecular characterization, x, 31, 38, 109, 111
molecular weight, 40, 52, 76, 77, 96
molecules, 8, 14, 15, 20, 29, 41, 42, 51, 52, 57, 71, 72, 73, 75, 81, 82, 83, 91, 92
morphological characterization, x, 32, 112
morphology, 23, 32, 85, 89
mutant, 64, 65, 95

N

neutral, 51, 54, 85, 121
nickel, 56, 88, 104, 119
nitrogen, 3, 52, 63, 64, 113, 118
nuclear magnetic resonance (NMR), 56, 75, 76, 92, 93, 94
nucleic acid, 16, 33, 42, 45, 47, 50, 53, 54, 82, 87
nutrients, 7, 14, 16, 21, 24, 25, 28, 31, 50, 52, 66, 86, 113, 114, 115, 146
nutrition, 16, 53, 93, 126

O

oil, 1, 2, 3, 4, 5, 7, 11, 12, 53, 54, 64, 80, 98, 101, 113, 118, 121
organ, ix, 60, 91
organic compounds, 14, 17, 24, 27, 57, 59
organism, v, ix, x, 4, 8, 13, 14, 15, 16, 17, 19, 23, 24, 25, 26, 27, 29, 31, 32, 36, 38, 42, 44, 45, 48, 55, 58, 64, 79, 86, 89, 91, 92, 93, 94, 95, 96, 105, 145
ox, 59, 62, 66, 74
oxidation, 5, 16, 26, 27, 55, 59, 60, 72, 73, 74, 80, 81, 82, 85, 89, 107, 116, 122
oxygen, 2, 6, 11, 12, 21, 25, 26, 55, 59, 60, 61, 66, 67, 68, 69, 74, 85, 86, 117, 118, 120, 121

P

passivation, x, 79, 80, 86
pathogens, 18, 19, 20, 32
pathway, 59, 67, 72, 93, 95
PCR, 31, 33, 34, 35, 36, 37, 38, 39, 40, 41, 42, 44, 45, 46, 89, 90, 108, 111, 112, 124, 125
peptides, 54, 58, 72, 73
peroxide, 59, 60, 61, 64, 68, 118, 121, 122
pH, 9, 10, 15, 27, 63, 85, 98, 101, 113, 120, 122
phenotype, 92, 93, 94, 126, 127
phosphate, 49, 54, 55, 64, 87, 114, 118
phosphorus, 1, 6, 7, 99, 119
physiology, 32, 45, 91, 107, 116, 121
pipeline, 2, 5, 11, 12, 19, 26, 64, 100
plants, 18, 21, 50, 51, 52, 58, 83, 94, 108
polar, 24, 81, 89
polarization, 7, 9, 61, 68, 82, 98, 121
polyacrylamide, 34, 38, 39, 41
polymerase, 36, 46, 110
polymers, 7, 16, 49, 50, 55, 57, 58, 60, 63, 115, 125
polymorphism, 35, 36, 39, 41, 110, 111, 112
polysaccharides, 7, 13, 15, 16, 20, 21, 49, 50, 51, 53, 54, 55, 57, 59, 69, 82, 112, 114, 115, 116

134 Index

population, 10, 21, 33, 39, 42, 47, 63, 82, 94, 100, 110, 119
power plants, 3, 14, 21
project, 105, 118, 147
prokaryotes, 24, 33, 83
protection, 10, 12, 53, 101, 119
proteins, vi, ix, 13, 15, 16, 20, 21, 42, 43, 47, 49, 50, 52, 53, 54, 55, 56, 57, 58, 59, 60, 61, 63, 68, 74, 91, 93, 96, 113, 114, 117, 122
proteome, 91, 92, 127
proteomics, x, 91, 92, 93, 94, 95, 96, 127
protons, 2, 6, 27, 85
Pseudomonas aeruginosa, 15, 20, 52, 82, 103, 105
pulp, 104, 106, 125

Q

quantification, 92, 94, 95, 96

R

randomly amplified polymorphic DNA (RAPD), 31, 36
reactions, ix, 2, 7, 10, 25, 49, 55, 58, 59, 60, 65, 66, 69, 71, 73, 74, 92, 122
repetitive extragenic palindromic-PCR (REP-PCR), 31, 37, 38, 111
research, ix, 3, 5, 66, 71, 77, 82, 91, 92, 93, 96, 98, 99, 102, 103, 104, 109, 118, 126, 145, 146, 147
researchers, 57, 59, 61, 64, 145
resistance, 11, 13, 14, 18, 66, 86, 103, 117, 123
resolution, 35, 37, 41, 42
resources, xi, 80, 93, 126, 127
respiration, 66, 119, 120, 121
response, 35, 54, 59, 72, 73
restriction fragment length polymorphis, 34, 110, 111, 112
reverse sample genome probing (RSGP), 31, 44
ribosomal Intergenic Spacer Analysis (RISA), 31, 38, 45
RNA, 24, 31, 42, 43, 93, 110, 112

S

Salmonella, 13, 19, 37, 52, 125
scanning electron microscopy (SEM), 6, 49, 56, 85, 88, 101
science, 57, 61, 91, 93, 96, 126, 145
sea water, 4, 79, 98, 104
sediments, 24, 28, 82, 83, 123
sequencing, 31, 36, 39, 40, 45, 46, 93, 106, 107, 109, 111, 112
shape, 20, 24, 32, 41, 47
signals, 44, 49, 94
silicon, 79, 80, 81, 82, 88, 124, 125
single-strand conformation polymorphism (SSCP), 31, 41, 45
sludge, 85, 87, 104, 106, 125
solubility, 51, 65, 71
solution, 6, 9, 73, 119, 121
solvent, 57, 71, 75, 76
species, 5, 10, 16, 17, 18, 19, 20, 24, 25, 26, 32, 34, 35, 36, 37, 38, 41, 42, 43, 44, 45, 47, 54, 55, 56, 59, 62, 65, 67, 79, 80, 82, 83, 95, 97, 104, 108, 110, 114, 118, 124, 125
stainless steel, 1, 3, 13, 18, 57, 59, 61, 86, 88, 101, 104, 117, 118, 119, 120, 121, 122
state, 6, 54, 58, 67, 74, 89, 92, 97, 113, 117
steels, ix, 1, 3, 6, 7, 8, 9, 10, 12, 13, 18, 26, 54, 55, 57, 59, 60, 61, 62, 63, 64, 65, 66, 88, 89, 97, 98, 99, 100, 101, 102, 104, 105, 106, 113, 114, 117, 118, 119, 120, 121, 122
sterile, 17, 88, 89, 90, 101
storage, 3, 18, 50, 92, 123
stress, 4, 94, 116
structure, 8, 13, 18, 20, 23, 24, 32, 33, 35, 36, 37, 38, 39, 40, 41, 45, 50, 51, 61, 63, 66, 75, 79, 86, 105, 108, 109, 110, 111, 112, 115, 116, 122, 123
substrates, x, 14, 18, 42, 44, 52, 54, 57, 58, 59, 62, 65, 66, 67, 87, 109
sulfate, 29, 53, 97, 98, 99, 100, 101, 113, 114, 117, 119, 121
sulfur, 1, 25, 27, 29, 64, 66, 100, 116, 121
sulphide, 1, 3, 4, 6, 7, 8, 9, 10, 11, 15, 63

Index

surface, x, 1, 2, 6, 7, 8, 9, 10, 13, 14, 15, 16, 17, 18, 19, 20, 21, 24, 26, 33, 46, 47, 49, 53, 54, 55, 57, 59, 61, 62, 64, 65, 66, 67, 68, 69, 79, 80, 81, 82, 83, 85, 86, 88, 89, 97, 102, 103, 104, 105, 118, 119, 120
surface charge density, x, 79, 81
survival, 14, 16, 23, 31, 83
synthesis, 51, 52, 58, 103

T

target, 33, 36, 41, 42, 44, 45, 87, 93
techniques, x, 24, 31, 32, 33, 34, 42, 43, 45, 49, 56, 68, 73, 75, 76, 77, 85, 91, 92, 95, 96, 101, 110, 112, 127
technology, ix, 44, 46, 47, 92, 93, 96, 98, 109, 127, 145, 146, 147
temperature, 2, 6, 24, 34, 39, 40, 41, 110
testing, 44, 47, 84, 94
transcriptomics, x, 91, 92, 94, 95, 96
transformation, 24, 27, 123, 124
transport, 17, 28, 91
treatment, 28, 52, 67, 82, 90, 102

U

uniform, 17, 40, 51

United States (USA), 11, 89, 90, 99, 100, 101, 104, 106, 110, 115, 146, 147

V

variations, 19, 34, 36, 39, 60

W

water, 3, 4, 6, 7, 8, 11, 13, 14, 16, 18, 19, 21, 26, 50, 51, 54, 59, 60, 61, 63, 65, 79, 80, 82, 87, 88, 90, 95, 97, 98, 100, 101, 102, 103, 104, 105, 107, 108, 120, 121, 125, 126, 147
weight, vii, 40, 52, 53, 76, 77, 83, 85, 88, 89, 96
worldwide, 1, 12, 71

X

x-ray, 56, 67, 85, 88

Y

yeast, 7, 64, 127
yield, 33, 43, 52, 73, 124

About the Authors

Santosh Kumar Karn
santoshkarn@gmail.com.

Dr. Santosh Kumar Karn is an Associate Professor in the Department of Biochemistry and Biotechnology at Sardar Bhagwan Singh University, Balawala, Dehradun, India. His science and technology research has been published in over 100 peer-reviewed articles and chapters, in addition to receiving significant media attention.

Professor Karn is an Associate Editor of the *Journal of Environmental Biology*, *Frontier in Microbiology* (Section: Microbiotechnology and Food Microbiology), *Frontier in Environmental Science*, and others. Professor Karn was the reviewing expert for FONDECYT Regular grant competition of the Chilean National Science and Technology Commission (CONICYT - Chile) and is involved in several national and international organizations. He deposited 100 different gene sequences to GeneBank-NCBI, 15 industrially important microorganism to National Collection of Industrial Microorganisms (NCIM, NCL, Pune, India) and five microorganisms to China General Microbiology Culture Collection Centre (CGMCC, Beijing).

Professor Karn received awards from DBT-JRF Award, Government of India, Prestigious International Young Scientist Award and Visiting Researcher Award through the Chinese Academy of Science (CAS), and the National Natural Science Foundation (NFSC) of China. He reviewed 225 original research papers and received the Publons Peer-Review Award (2017 and 2018) as one of the top 1% peer reviewers in microbiology and editorial pursuit internationally by Clarivate Analytics, USA. He also received the award from Elsevier and Springer. In the year 2016 he received the award for Presidential International Fellowship Initiative (PIFI initiative) by CAS Beijing, China.

Professor Karn ventures to inspire students with his enthusiasm for the subjects, cultivate critical-thinking and decision-making for innovative approaches.

Anne Bhambri
anne.bhambri26@gmail.com.

Anne Bhambri has an M.Sc. in Biochemistry from Sardar Bhagwan Singh University, Dehradun, and received Silver Medal at the post-graduate level and is currently a doctoral student. She works on biological nutrient removal using specific microbes guided by S. K. Karn in the Department of Biochemistry & Biotechnology at S. B. S. University, Dehradun, India.

To date, Ms. Bhambri has published 20 research papers in leading international journals: *Scientific Reports* (Nature's Publishers), London, UK; Elsevier, Netherland; Nova Science Publishers, USA; *Corrosion Review*, Germany; Springer, Germany; *Chemistry & Ecology* (Taylor & Francis Group), London; Wiley-Scrivener, Chichester, UK; Bentham Science, UAE; *Geomicrobiology Journal* (Taylor & Francis Publishers), United Kingdom etc.). She has conducted two proceedings, one in Emerging Scientist-ES2022 organized by Asian Council of Science Editors jointly with Academy of Excellence in Bioscience, and the second in the National Conference organized by National Institute of Technology, Warangal (2021). Ms. Bhambri presented a paper in the National Conference and received

"Best Oral Presentation Award" from the National Institute of Technology, Warangal (2021) and "Best P.G. Thesis Award" in another National Conference Oral Presentation (2019).

Currently Ms. Bhambri is working as a project associate in the research project sponsored by M/s Xcel Life Science, California, USA. Recently in (2022) she received Young Women Scientist Excellence Award by USERC, Dehradun, Government of Uttarakhand for her contribution in Science & Technology. Recently she has published work in *Frontier in Microbiology* and edited two books with Nova Science Publisher, USA.

Ms. Bhambri has broad interests in soil and water pollution and their remediation, medicinal mushroom cultivation and phytochemical extraction, environmental biotechnology and nanotechnology, bioremediation, and more.